Reuse of Sludge and Minor Wastewater Residuals

Alice B. Outwater

Library of Congress Cataloging-in-Publication Data

Outwater, Alice B.
Reuse of sludge and minor wastewater residuals/Alice B. Outwater.
p. cm.
Includes bibliographical references and index.
ISBN 0-87371-677-9
1. Sewage sludge—Recycling. 2. Title.
TD774.09 1994
628.3′8—dc20 93-50138
CIP

Lewis Publishers is an imprint of CRC Press

International Standard Book Number 0-87371-677-9
Library of Congress Card Number 93-50138
Printed in the United States of America 1 2 3 4 5 6 7 8 9 0
Printed on acid-free paper

AUTHOR

Alice B. Outwater has a B.S. in Mechanical Engineering from the University of Vermont (1983) and an M.S. in Technology and Policy from Massachusetts Institute of Technology (1987). She worked in residuals management for the Massachusetts Water Resources Authority, which managed the largest wastewater engineering project of the decade: the $6 billion Boston Harbor Clean-Up. As an engineer specializing in minor residuals and innovative technologies for sludge reuse, Ms. Outwater was part of the team that planned and implemented the program that shifted 200 dry tons of sludge a day from ocean disposal to beneficial reuse. She has written numerous technical papers on sludge and scum reuse. This book is a practical handbook for the reuse of sludge and minor residuals.

CONTENTS

Reuse of Sludge and Minor Wastewater Residuals

1 INTRODUCTION

The land application of sludge is as old as agriculture. The Chinese have used human waste as a fertilizer for thousands of years. The Japanese called it nightsoil and carried it out of town under the cover of darkness to apply to their orchards and gardens. The special role of the Untouchable caste in India was to gather up human waste from the towns and transport it out to the countryside. Before the 1940s, land application of sludge was common in the United States. In those simpler, less squeamish times the "manurial" benefits of sludge were appreciated, and local farmers were glad to use wastewater residuals to improve crop yields. However, as soon as synthetic fertilizers became affordable and available, sludge was seen as a disposal problem rather than a resource.

After a hiatus of nearly 50 years, the amount of sludge that is applied to U.S. forests, crops, and disturbed land is steadily increasing. Seattle sprays sludge on forests to promote tree growth, Maryland sludge compost is used to enhance the turf on the White House lawn, Omaha spreads its sludge on nearby farmland, Pennsylvania rehabilitates strip-mined land with sludge compost, and Florida uses sludge from around the country to fertilize the orange groves. Nearly all of the major cities in the U.S., including Philadelphia, Chicago, Baltimore, Denver, Madison, Phoenix, Los Angeles, and Portland, are processing their municipal sludge into fertilizers and soil conditioners.

When sludge products are used judiciously, the risk to human health and the environment is negligible. There are rarely reports of harmful effects to humans, livestock, and pets or to the environment resulting from the land application of sludge products when recommended application rates and good management practices are followed, in spite of the fact that we live in a litigious society and sludge has been land applied for generations. With the promulgation of the U.S. Environmental Protection Agency (EPA) 40 CFR Part 503 regulations,[1] the beneficial reuse of sludge has become a viable alternative for almost all wastewater treatment plants in the U.S.

1.1 SLUDGE PRODUCTION IN THE UNITED STATES

There is a tremendous amount of sludge produced in this country. A typical family of four discharges 300 to 400 gal of wastewater per day, using 75 to 100 gal per person. Domestic wastewater contains material flushed into household drains from sinks, toilets, and tubs. Components of domestic sewage include soaps, shampoo, human excrement, toilet paper, foodstuffs, detergents, pesticides, household hazardous waste, and oil and grease. Domestic wastewater may be treated (or partially treated) at its source in septic tanks, cesspools, or portable toilets or in publicly or privately owned treatment works. These treatment works may accept domestic wastewater alone or a combination of domestic and industrial wastewater.

Municipal wastewater treatment works may use one or more levels of treatment — primary, secondary, or tertiary — to clean this wastewater. Each level of treatment provides both a higher level of wastewater cleanup and a larger quantity of sewage sludge. Unprocessed sewage sludge contains from 93 to 99.5% water, as well as the solids and dissolved substances that were present in the wastewater or that were added or cultured by the wastewater treatment process.

Approximately 12,750 publicly owned treatment works (POTWs) generate 5.4 million dry metric tons of sludge annually, or 47 dry pounds of sewage sludge for every person in the U.S. This sewage sludge is commonly used or disposed of in a number of ways. Sludge is applied to agricultural and nonagricultural lands; it is sold or given away for use in home gardens, lawns, horticulture, and landscaping (often referred to as sludge distribution and marketing); it is disposed of in municipal landfills, sludge-only landfills (known as monofills), and surface disposal sites; and it is incinerated.

Table 1-1 shows the amount of sludge generated based on the size of the facility and on the amount of sewage sludge that is disposed of by a use or disposal practice. Table 1-2 shows the number of facilities employing a particular method of use or disposal. Only about one third of the sewage sludge generated in the U.S. is effectively reused by land application or is sold or given away for use in home gardens.[2] In comparison, Japan uses 42% of its sewage sludge for coastal reclamation and home gardening or farming uses, while the U.K. applies 51% of its sewage sludge to the land.

1.2 BENEFITS OF REUSING SEWAGE SLUDGE

Sludge reuse improves the productivity of the land by adding organic matter and nutrients to the soil. In addition, there are a number of indirect advantages to human health and the environment. Secondary benefits of sludge reuse include a reduction of the adverse health effects of incineration, a decreased dependence on chemical fertilizers, a reduction in the emissions associated with incineration that may contribute to the greenhouse effect, and a reduction in fuel or energy costs associated with incineration. EPA analyses show that the land application of sludge poses less carcinogenic risk than sludge incineration. The lifetime cancer risk to a highly exposed individual ranges from 6×10^{-4} for land application and surface disposal of sludge to 7×10^{-3} for incineration.

The organic and nutrient contents of sewage sludge are a valuable resource for improving marginal lands and as a supplement to fertilizers and soil conditioners. Sludge use as a fertilizer can both increase crop yields and decrease the farmer's costs for chemical fertilizers. Studies done across the country have shown that the benefits to farmers that reuse this resource extend beyond avoiding the costs of chemical

Table 1-1 Estimated Mass of Sludge Disposed Annually in the U.S.[a]

	Reported flow rate (MGD)[b]				
Use or disposal practice	**>100**	**>10 to 100**	**>1 to 10**	**≤1**	**Total[a]**
Incineration	382.9	346.5	124.8	10.5	864.7
Land application: agricultural	203.0	400.8	423.9	143.2	1170.9
Land application: compost	22.4	65.3	31.7	30.8	150.2
Land application: forests	4.5	24.5	1.0	1.3	32.3
Land application: public contact	62.1	60.5	40.3	6.3	166.1
Land application: reclamation	52.6	9.8	2.4	1.0	65.8
Land application: sale	30.6	27.8	11.9	0.8	71.1
Land application: undefined	12.7	76.4	27.2	13.0	129.3
Co-disposal: landfill	518.6	674.0	495.6	110.4	1818.7
Surface disposal: dedicated site	34.2	124.9	63.2	36.5	258.8
Surface disposal: monofill	13.8	79.8	41.6	22.2	157.4
Surface disposal: other	31.5	60.0	17.4	28.5	137.5
Unknown: ocean[c]	166.1	157.9	8.0	3.4	335.5
Total	1532.0	2128.3	1284.1	407.7	5357.2

Source: U.S. EPA.[2]

[a] Mass values are in dry metric tons $\times 10^3$.

[b] MGD = millions of gallons per day.

[c] This survey was conducted before the Ocean Dumping Ban Act of 1988, which generally prohibited the dumping of sewage sludge into the ocean after December 31, 1991. Ocean dumping of sewage sludge ended in June 1992. Numbers may not add up to 100% because of rounding.

Table 1-2 Number of POTWs and Sludge Quantities by Use or Disposal Practice

	POTWs use or disposal practice		Quantity of sludge disposed[a]	
Use/disposal practice	**Number of POTWs**	**Percent of POTWs**	**Quantity[b]**	**Percent of sludge**
Incineration	381	2.8	864.7	16.1
Land application	4,657	34.6	1,785.3	33.3
Codisposal: landfill	2,991	22.2	1,818.7	33.9
Surface disposal	1,351	10.0	553.7	10.3
Unknown: ocean disposal[c]	133	1.0	335.5	6.3
Unknown: other	3,920	29.1	0.0	0.0
Unknown: transfer	25	0.2	N/A	N/A
All POTWs	13,458	100.0	5,357.2	100.0

Source: U.S. EPA.[2]

[a] N/A = not applicable.

[b] In dry metric tons $\times 10^3$.

[c] This survey was conducted before the Ocean Dumping Ban Act of 1988, which generally prohibited the dumping of sewage sludge into the ocean after December 31, 1991. Ocean dumping of sewage sludge ended in June 1992.

fertilizers. The increased organic content in sludge-amended soil improves water retention and drainage, lightens heavy soils, and enriches sandy soils. Depending on crop rotation, previous soil management practices, soil type, and the level of sludge application, farmers can enjoy substantial financial benefits from this resource. About two thirds of the sludge that is land applied in this country is used to improve the condition and the nutrient content of the soil for agricultural crops, including row and feed crops and pastures.

The beneficial uses of sludge are not restricted to agricultural lands; about 22% of the sludge that is land applied is used on nonagricultural sites. Sewage sludge is used in silviculture to increase forest productivity and to revegetate and stabilize harvested forestland. The application of sludge to forestland shortens the wood production cycle by accelerating tree growth, particularly on marginal soils. Studies from the University of Washington have shown that trees grow twice as fast on sludge-amended soil. This means that trees that would typically be cut after 60 years could be cut after only 30 years, radically improving the economic forecasts for these projects. Forest soils are well suited to sludge application because they have high rates of infiltration to reduce ponding and runoff, large amounts of organic material to immobilize metals in the sludge, and perennial roots systems which allow year-round application in mild climates.

Sludge is used to stabilize and revegetate areas destroyed by natural disasters like forest fires or landslides and by human activities such as mining, dredging, and construction. Historically, land reclamation with sludge application has been very successful and somewhat less expensive than other methods of land reclamation. Philadelphia sludge has been used to reclaim over 3000 acres of strip-mined lands. In a strip-mined area in Fulton County, IL, reclamation with sewage sludge cost $3660 an acre, while reclamation using commercial methods cost $3395 to $6290 an acre. Studies on Pennsylvania strip mines have shown that the microbial communities in sludge-reclaimed land revert to those of normal soils within 2 or 3 years, while the land reclaimed with conventional methods can take 10 to 15 years and even longer.

Sludge is used to fertilize highway median strips, cloverleaf exchanges, airports, parks, ball fields, cemeteries, and other public access sites and to cover closed landfills. Using sludge as an aid in establishing the final vegetative cover for municipal solid waste landfill takes advantage of the nutrient content and soil amendment characteristics of sewage sludge. The quick establishment of the final root zone can facilitate municipal solid waste closure plans.

Finally, sludge can be processed, packaged, and sold or given away for use in home gardens. About 12% of the sewage sludge generated in the U.S. is distributed and marketed as a soil amendment or fertilizer.

1.3 CONSTRAINTS FOR SEWAGE SLUDGE REUSE

To beneficially reuse sewage sludge, the standards laid out in the 40 CFR Part 503 regulations must be met. The benefits of using sludge to improve land productivity are substantial, but if the sludge contains high levels of pathogens or high concentrations of pollutants, land application could contaminate the soil, water, crops, livestock, fish, and shellfish. Virtually all sewage sludge contains nutrients such as nitrogen and phosphorus, but it also contains significant numbers of pathogens, including bacteria, viruses, protozoa, and eggs of parasitic worms. In addition, some sludges contain more than trace amounts of volatile and semivolatile organic compounds and heavy metals along with the organic constituents. The pollutants come from domestic wastewater, from the discharge of industrial wastewater to municipal sewers, and from combined sewers which collect the runoff from streets, parking lots, lawns, and fields where fertilizers and pesticides were applied incorrectly.

The major threats to human health and the environment in the land application of sludge are related to pathogens, metals, and persistent organic chemicals content. The 40 CFR Part 503 regulations are designed to prevent the contamination of soil and crops by pathogens. They also ensure that food and animal feed crops are not contaminated by inorganic chemicals in the sludge. The Part 503 regulations are based on a multimedia approach which ensures that the land application of sludge presents a negligible risk to human health and the environment.

1.3.1 Improving Sludge Quality

If the levels of pollutants in the sludge exceed the maximum standards provided by the Part 503 rule, sludge quality must be improved before it can be reused. The most cost-effective method of removing toxic contaminants from the sludge is to prevent them from entering the wastewater system in the first place. POTWs with capacities greater than 5 million gallons per day (MGD) are required to establish discharge limits for industrial dischargers to the wastewater system by 40 CFR 403, a regulation which was initially disregarded by many POTWs. In the last decade, pretreatment programs have sprung up around the country, and 2015 of the nation's POTWs currently have local pretreatment programs. While this is only about 15% of the total number of POTWs in the country, it includes all of the largest facilities and is estimated to be about 80% of the national wastewater flow. In many major cities, the level of industrial toxics discharged to the wastewater system has quietly decreased by tenfold or more in the past decade. In most cases, improving sludge quality is a relatively straightforward problem to resolve.

1.4 A GUIDE TO SLUDGE REUSE

This book is a guide to sludge reuse. It provides an overview of the production and characterization of sludge and minor residuals and presents a practical, least cost approach to improving sludge quality. Markets are paired with sludge products, and the problems and solutions of siting a sludge processing facility are explored. Moving from the simpler to the more complex sludge processing technologies, case histories give a practical view of the costs and the caveats of each sludge management option.

Starting with the land application of liquid sludge on agricultural land and forestland, we move to natural and mechanical dewatering methods and sludge cake application. Next, lime stabilization, pasteurization, and chemical fixation are discussed. Sludge composting processes from static pile to in-vessel composting are included, as are the odor control options for composting facilities and the markets for sludge compost. Sludge pelletization and some innovative technologies for sludge reuse are presented, as well as an analysis of the new EPA 40 CFR Part 503 regulations.

REFERENCES

1. "40 CFR Part 503 Regulations," U.S. Environmental Protection Agency (1993).
2. "National Sewage Sludge Survey," U.S. Environmental Protection Agency, (1988).

2 SLUDGE

Sludge quantity and quality are the two key factors in any sludge reuse program. The first task of a beneficial reuse program is to quantify these variables thoroughly.

2.1 SLUDGE CHARACTERIZATION

Sludge which is land applied must comply with federal and state regulations that restrict the levels of contaminants in order to protect the public health and the environment. Since the options for beneficial reuse will be determined in part by the level of contaminants in the sludge, a sampling program that accurately reflects sludge quality is a critical step on the path toward beneficial reuse.

Sludge quality measurements are usually made of the parameters listed in Table 2-1 on a monthly, semiannual, or annual basis. The standard parameters provide a measure of the fertilizer value of the sludge, while the priority pollutants (and metal levels in particular) give the concentrations of contaminants which may restrict or preclude the possibility of land application.

2.1.1 Physical Composition

The characteristics of sludge[1] depend on the origin, the amount of aging that has taken place, and the type of processing the sludge has received. Sludge from primary sedimentation tanks is usually gray and slimy and, in most cases, smells awful. It can be readily digested under suitable conditions.

Sludge from chemical precipitation tanks is usually black, although its surface may be red if it contains much iron. Its odor may be objectionable, but it doesn't smell as bad as primary sludge. It is somewhat slimy, but the hydrate of iron or aluminum in it makes it gelatinous. If it is left in the tank, it undergoes decomposition at a slower rate than primary sludge. It gives off gas in substantial quantities and its density is increased by standing.

Activated sludge is generally brown and flocculent. If the color is dark, it may be approaching septic conditions. If the color is lighter than usual it may have been underaerated, with a tendency for the solids to settle more slowly. When activated sludge is in good condition its odor isn't too bad, but it tends to becomes septic rather quickly, in which case it stinks of putrefaction. Activated sludge can be digested alone or mixed with other fresh sewage solids.

Table 2-1 Standard Sludge Analyses

Standard parameters	Priority pollutants
pH	Metals and other inorganics
Chlorides	Pesticides
% Volatile solids	Halogenated aliphatic hydrocarbons
% Total solids	Monocyclic aromatic hydrocarbons
Total Kjeldahl nitrogen	Polycyclic aromatic hydrocarbons
Ammonia	Halogenated ethers
Nitrates	Phthalate esters
Total phosphorus	Polychlorinated biphenyls and related compounds
Available phosphorus	Nitrosamines and other N compounds
Potassium	
TCLP	
SOUR	

Source: U.S. EPA.[5]

Trickling filter humus is brownish, flocculent, and quite innocuous when fresh. It decomposes more slowly than other undigested sludges unless it is full of worms. Wormy humus is truly foul, though readily digested.

Digested sludge is dark brown to black and contains exceptionally large quantities of gas. When thoroughly digested, its odor is relatively faint and reminiscent of hot tar, burnt rubber, or sealing wax. When drawn off in porous beds in layers 6 to 10 in. deep, the solids are first carried to the surface by the entrained gas, leaving a comparatively clear sheet of water below which drains off rapidly, allowing the solids to sink down slowly onto the bed. As the bed dries, the gas escapes, leaving a well-cracked surface with an odor of garden loam.

Sludge from septic tanks is black and, unless well digested by long storage, stinks of hydrogen sulfides and other gases.

2.1.2 Chemical Composition

Table 2-2 gives some typical data on the chemical composition of raw and digested sludge. Nutrient levels are important for the beneficial reuse of sludge, and the fertilizer value is based primarily on the content of nitrogen, phosphorus, and potash. The pH, alkalinity, and organic content are important in process control of anaerobic digestion.

In the late 1980s, the EPA studied sludge from 209 randomly selected sewerage systems from all regions of the U.S. A wide range of plant sizes were sampled and analyzed in three broad categories: classicals, metals, and organic compounds. Classical analytes include total Kjeldahl nitrogen (TKN), phosphorus, and cyanide. The survey analyzed 69 metals from aluminum to zirconium and over 330 organic compounds, including pesticides, herbicides, dioxins/furans, volatile and semivolatile organic compounds, and polychlorinated biphenyls (PCBs).[2] Table 2-3 provides a summary of the results for regulated pollutants.

Most organic compounds were not detected with any regularity. Samples were analyzed on a wet weight basis, so detection limits were not uniform and varied based on the percent total solids of the sludge sample.

The results from the National Sewage Sludge Survey (NSSS) are compared to the results from the EPA 40 City Survey[3] completed in 1978 (Table 2-4). The levels of contaminants in sludge fell considerably in the decade between the 40 city survey and

Table 2-2 Typical Chemical Composition of Raw and Digested Sludge

	Raw primary sludge		Digested sludge	
Item	**Range**	**Typical**	**Range**	**Typical**
Total dry solids (TS), %	2.0–7.0	4.0	6.0–12.0	10.0
Volatile solids (VS), %	60–80	65	30–60	40.0
Grease and fats (ether soluble, % of TS)	6.0–30.0	—	5.0–20.0	
Protein (% of TS)	23–30	25	15–20	18
Nitrogen (N, % of TS)	1.5–4.0	2.5	1.6–6.0	3.0
Phosphorus (P_2O_5, % of TS)	0.8–2.8	1.6	1.5–4.0	2.5
Potash (K_2O, % of TS)	0–1.0	0.4	0.0–3.0	1.0
Cellulose (% of TS)	8.0–15.0	10.0	8.0–15.0	10.0
Iron (not as sulfide)	2.0–4.0	2.5	3.0–8.0	4.0
Silica (SiO_2, % of TS)	15.0–20.0	—	10.0–20.0	—
pH	5.0–8.0	6.0	6.5–7.5	7.0
Alkalinity (mg/l as $CaCo_3$)	500–1500	600	2500–3500	3000
Organic acids (mg/l as HAc)	200–2000	500	100–600	200

Source: Metcalf & Eddy, Inc., *Wastewater Engineering: Treatment, Disposal, and Reuse,* McGraw-Hill, 1972. With permission.

Table 2-3 Summary of Concentrations for Regulated Pollutants

Pollutant	Number of times detected	Mean (mg/kg)	Minimum (mg/kg)	Maximum (mg/kg)
4,4′-DDD	1	0.391	0.391	0.391
4,4′-DDE	4	0.100	0.030	0.190
4,4′-DDT	7	0.051	0.015	0.121
Aldrin	8	0.029	0.019	0.046
Arsenic	194	12.390	0.300	315.600
Benzene	4	0.098	0.012	0.220
Benzo*(a)*pyrene	7	10.785	0.671	24.703
Beryllium	64	0.660	0.100	3.900
Bis(2-ethylhexyl)phthalate	189	107.233	0.510	89.129
Cadmium	194	65.460	0.700	8,220.000
Chlordane	1	0.489	0.489	0.489
Chromium	231	258.515	2.000	3,750.000
Copper	239	665.300	6.800	3,120.000
Dieldrin	6	0.024	0.013	0.047
Dimethyl nitrosamine	0	—	—	—
Heptachlor	1	0.023	0.023	0.023
Hexachlorobenzene	0	—	—	—
Hexachlorobutadiene	0	—	—	—
Lead	213	195.230	9.400	1,670.000
Lindane (gamma-BHC)	2	0.074	0.072	0.076
Mercury	184	4.120	0.200	47.000
Molybdenum	148	13.120	2.000	67.900
Nickel	201	77.010	2.000	976.000
PCB-1016	0	—	—	—
PCB-1221	0	—	—	—
PCB-1232	0	—	—	—
PCB-1242	0	—	—	—
PCB-1248	23	0.740	0.043	5.203
PCB-1254	13	1.765	0.312	9.347
PCB-1260	20	0.671	0.031	4.006
Selenium	163	6.240	0.500	70.000
Toxaphene	0	—	—	—
Trichloroethylene	7	0.848	0.024	3.302
Zinc	239	1,692.760	37.800	68,000.000

Source: U.S. EPA.[5]

Table 2-4 National Sewage Sludge Survey (NSSS) and 40 City Survey Results

Pollutant	NSSS mean (mg/kg)	40 city mean (mg/kg)
Cadmium	65.46	120
Chromium	258.515	480
Copper	665.3	710
Lead	195.23	370
Mercury	4.12	3
Nickel	77.01	140
Zinc	1692.76	1860

Source: Kuchenrither and McMillan.[2]

the NSSS. Levels of cadmium, chromium, lead, and nickel dropped by roughly 50%, while copper and zinc levels stayed roughly the same. Mercury levels appear to have risen, but this may be an artifact of improved analytical techniques.

Data from the NSSS indicate that levels of metals in cities that reuse their sludge are typically lower than the national averages. It is safe to assume that these cities did not opt to reuse their sludge because of low metals levels. Instead, it is evident that a beneficial reuse program goes hand in hand with a good sampling program and an aggressive program to monitor and reduce industrial contributions.

2.2 SLUDGE PRODUCTION

The basic wastewater treatment process has changed little in the past 50 years. Wastewater treatment plants are enormously capital-intensive facilities that last for generations. With so much capital at stake and within a regulatory framework that makes failure unacceptable, municipalities are unwilling to take the risk of implementing an untried technology. This means that in almost all cities across the U.S., wastewater treatment technologies and the sludge they produce are known quantities. Figure 2-1 is a typical schematic of a wastewater treatment process used across the country.

The influent flow at a treatment plant carries a wide range of debris along with the wastewater solids. Preliminary screening protects the plant equipment from debris. From the screens, the wastewater travels through grit chambers which slow the flow enough to allow the heavier inorganic materials such as sand and gravel to settle to the bottom. Grit and screenings are usually landfilled.

After preliminary treatment the flow enters the primary sedimentation tanks or clarifiers. Sedimentation tanks are designed to hold wastewater for 1 or 2 h. During that time, most of the heavier suspended solids settle to the bottom of the tank, where plough-like scrapers move the solids to a sump or hopper to be sucked out of the bottom of the tank. Primary wastewater treatment processes remove the solids that settle out of the wastewater by gravity. This generates 2500 to 3500 l of sludge per million liters of wastewater treated. Primary sludge contains 3 to 7% solids, 60 to 80% of which is organic matter. The water content of primary sludge can be reduced easily by thickening or dewatering.

The effluent from the primary basins still contains some suspended, colloidal, and dissolved solids. Secondary treatment promotes the growth of millions of

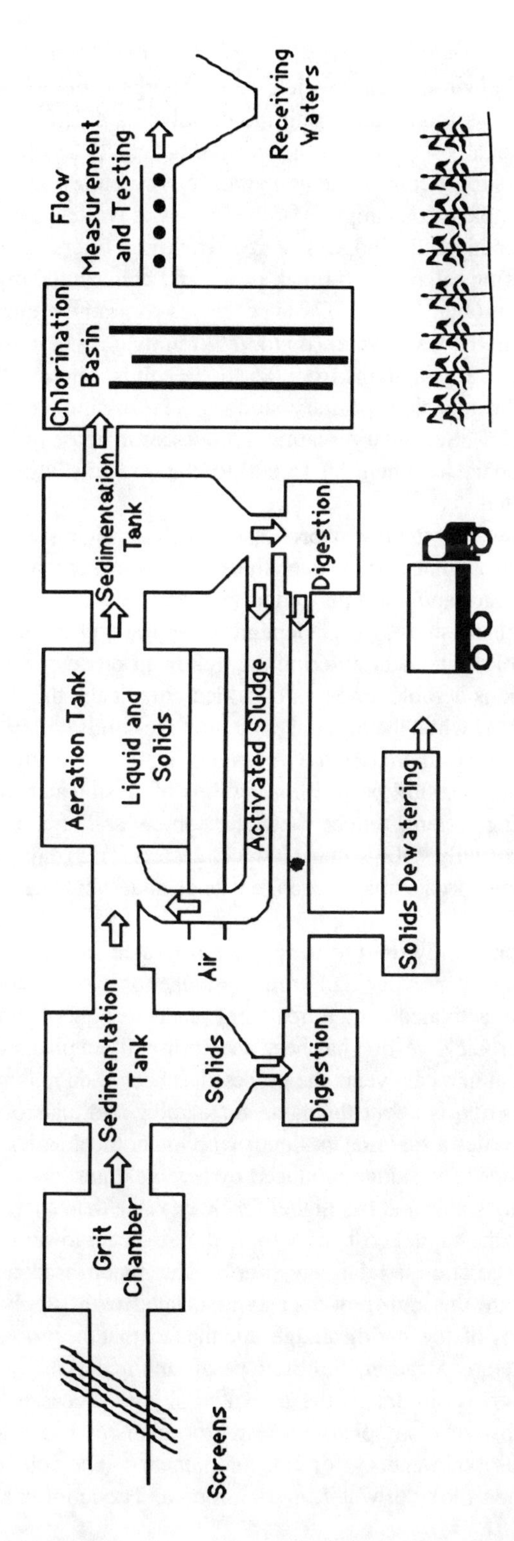

Figure 2-1 Wastewater treatment in the U.S. (Courtesy of MWRA)

microorganisms to consume the wastes present in the water. It creates a highly controlled artificial environment in which the microorganisms are encouraged to use dissolved pollutants as food, converting the dissolved solids into suspended solids (biomass) that physically settle out of the water at the end of secondary treatment. To succeed, the process must control the temperature, oxygen level, and contact time to allow rapid and complete consumption of the dissolved wastes. Secondary treatment processes include activated sludge systems, trickling filters, and other attached growth systems, using microbes to break down and convert the organic substances in wastewater to microbial residue. These processes commonly remove up to 90% of the organic matter in the wastewater (up to 95% removal is not unheard of), producing a sludge that typically contains from 0.5 to 2% solids. These solids are generally more difficult to dewater than primary sludges. The organic content of the solids ranges from 50 to 60%. Secondary treatment processes increase the volume of sludge generated over primary treatment by 15,000 to 20,000 l of sludge per million liters of wastewater treated.

Advanced wastewater treatment processes, such as chemical precipitation and filtration, produce an advanced or tertiary sludge. Chemical precipitation uses chemicals to remove organics and nutrients and to separate the solids from the wastewater. Characteristics of these sludges vary depending on the type of advanced treatment used and the type of wastewater entering the treatment process. Since these sludges typically contain considerable amounts of added chemicals, the solids content will vary from 0.2 to 1.5%, while the organic content of the solids will be in the 35 to 50% range. Tertiary treatment increases the volume of sludge generated over secondary treatment by another 10,000 l per million gallons of wastewater treated.

Sludge thickening systems reduce the water content and volume of sludge. From there, the sludge is usually stabilized in sludge digesters for 20 days or more to reduce odor and pathogenic organisms. A conventional anaerobic digester is shown in Figure 2-2.

Aerobic digestion is only used to treat waste-activated sludge, mixtures of waste-activated or trickling filter sludge and primary sludge, or waste sludge from extended aeration plants or in activated sludge treatment plants designed without primary settling. In the past, aerobic digestion has been used primarily in plants with capacities of less than 5 MGD, but in recent years the process has been used in larger facilities. The volatile solids reduction is about the same for aerobic and anaerobic digestion, but aerobic digestion creates a supernatant liquor with lower biochemical oxygen demand (BOD) concentrations. The sludge produced by aerobic digestion is odorless, humus-like, and biologically stable and has higher fertilizer value than anaerobically digested sludge. In addition, the capital costs of aerobic digestion are lower than for anaerobic digestion, and the operation is relatively simple. The major disadvantages of aerobic sludge processing are the high power costs associated with supplying the requisite oxygen, the difficulty of dewatering sludge, and the fact that the process is significantly affected by temperature, location, and the type of tank material.[1]

Aerobic digestion is similar to the activated sludge process. As the supply of available food (substrate) is depleted, the microorganisms begin to consume their own protoplasm to obtain energy for cell maintenance. The cell tissue is oxidized aerobically to carbon dioxide, water, and ammonia. The ammonia is subsequently

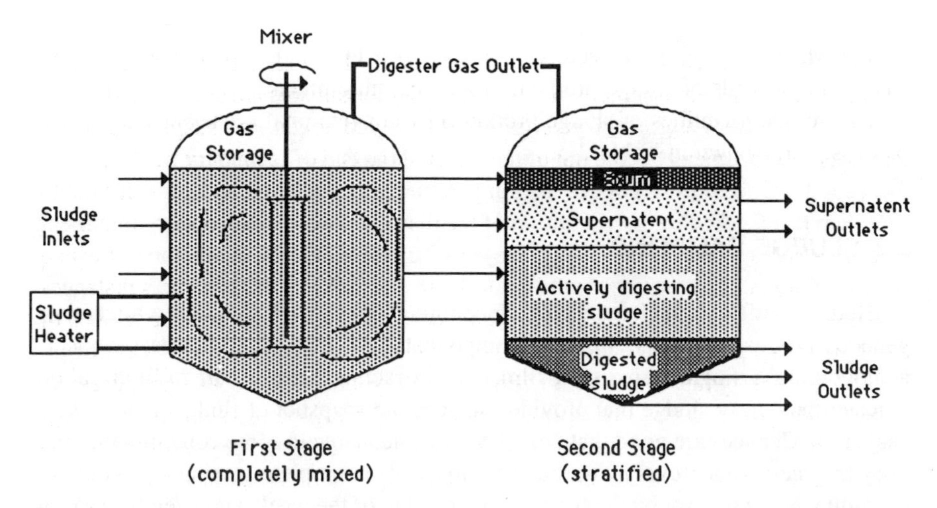

Figure 2-2 Schematic of a conventional two-stage anaerobic digester. (Redrawn from Metcalf & Eddy, Inc., *Wastewater Engineering: Treatment, Disposal, and Reuse,* McGraw-Hill, 1972. With permission.)

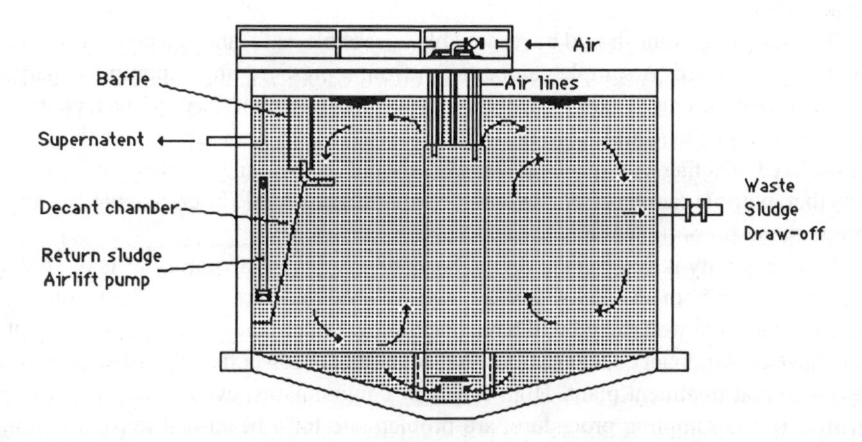

Figure 2-3 Schematic of a conventional aerobic digester. (Redrawn from Metcalf & Eddy, Inc., *Wastewater Engineering: Treatment, Disposal, and Reuse,* McGraw-Hill, 1972. With permission.)

oxidized to nitrate as digestion proceeds, creating a sludge with a relatively high fertilizer value. A conventional aerobic digester is shown in Figure 2-3.

In some areas of the U.S., digested sludge has been wasted to sea, landfilled, or incinerated. In the 1990s, sea disposal of sludge in no longer possible, landfill pressures are increasing, and incinerators are unpopular facilities. The result is a policy shift toward the beneficial reuse of sludge.

To design a sludge reuse program, the quantities and characteristics of the sludge to be processed must be clearly defined. The quantity of sludge generated by a treatment plant fluctuates over a wide range. If a POTW has been wasting its sludge to sea, there may be a good deal of uncertainty surrounding the exact amount of sludge which must be processed. Conversely, if a POTW has been landfilling its sludge, the records of sludge production may be quite precise.

The normal quantities of sludge produced by conventional wastewater treatment processes are shown in Table 2-5.

2.3 SLUDGE SAMPLING

Sludge is difficult to characterize, since municipal sewage sludge is a nonhomogeneous mixture of many different components. An individual sludge sample is analogous to a single frame of a film, and extracting from a half million gallon digester the pint of sludge that provides an accurate snapshot of sludge quality is no easy trick. Chances are good that any given sample is low in some contaminants and high in others, since every digester has areas of incomplete mixing. In addition, contaminant levels vary by the time of day, the day of the week, and even the season of the year. In systems with combined storm and sanitary sewers, rainfall will affect levels of certain contaminants, as will dry weather. The result is that most collections of sludge data from a single well-mixed digester have a broad range of contaminant concentrations.

The sampling point should be reasonably accessible, safe, and at a point where the flow is well mixed. A sample can be taken from a pressure line, a pump discharge, an open tank, a closed tank, a flotation tank, multiple containers, or a lagoon. In general, composite samples are more representative than grab samples. It should be considered whether a composite sample should be volume or time related and whether multiple samples from different waste streams should be composited according to some proportion.[4]

Sludge quality is determined on the basis of a series of laboratory tests of sludge samples. Often both an average and a range are used to determine the sludge quality. Sludge that is not well mixed will have pockets of contaminants, so any given sample will have contaminant concentrations that are higher or lower than the average sludge levels in that treatment plant. High levels of contaminants, even if they're only an artifact of the sampling procedure, are problematic for a beneficial reuse program. The purpose of developing a good sludge sampling program is to have lower contaminant concentrations, since well-mixed sludge doesn't have the slugs of contaminants. A thoughtful sampling program can be used to reduce much of the uncertainty and the spread of the values of contaminant concentrations.

One of the most intractable problems in sludge quality measurements is how to address outlying values. If some of the measurements of contaminant levels are significantly higher or lower than most other values, including these outliers in a cumulative average can significantly affect sludge quality estimates. If outlying values are assumed to be a valid representation of slug loadings from industrial contributors, improvements in sludge quality would be in the province of the industrial inspection team, and tactics may include increased industrial surveillance and reductions in permitted discharge levels or in the number of permits issued. Outlying values also may be artifacts of poorly chosen sampling ports or slipshod sampling or laboratory procedures. Refining the sampling program, through

Table 2-5 Normal Quantities of Sludge Produced by Different Treatment Processes

Treatment process	Quantity of sludge (gal/million gal of sewage)	Water content (%)	Specific gravity of sludge solids	Specific gravity of sludge	Dry solids (lb/million gal of sewage)
Primary sedimentation					
Undigested	2,950	95	1.40	1.02	1,250
Digested in separate tanks	1,450	94	—	1.03	750
Digested and dewatered on sand beds	—	60	—	—	750
Digested and dewatered by vacuum filters	—	72.5	—	1.0	750
Digested and dewatered by belt presses	—	72	—	1.0	750
Digested and dewatered by centrifuges	—	72	—	1.0	750
Trickling filter	745	92.5	1.33	1.025	476
Chemical flocculation	5,120	92.5	1.93	1.03	3,330
Dewatered on vacuum filters	—	72.5	—	—	3,300
Primary sedimentation and activated sludge					
Undigested	6,900	96	—	1.02	2,340
Undigested and dewatered on vacuum filters	1,480	80	—	0.95	2,340
Digested in separate tanks	2,700	94	—	1.03	1,400
Digested and dewatered on sand beds	—	60	—	—	1,400
Digested and dewatered on vacuum filters	—	80	—	0.95	1,400
Activated sludge					
Wet sludge	19,400	98.5	1.25	1.005	2,250
Dewatered on vacuum filters	—	80	—	0.95	2,250
Dried by heat driers	—	4	—	1.25	2,250
Septic tanks (digested)	900	90	1.40	1.04	810
Imhoff tanks (digested)	500	85	1.27	1.04	690

Source: Metcalf & Eddy, Inc., *Wastewater Engineering: Treatment, Disposal, and Reuse,* McGraw-Hill, 1972. With permission.

the analytical approach described in this chapter, may provide dramatic improvements in sludge quality.

2.3.1 Digester Stratification

Digesters are large enough containers that significant stratification is inevitable. Even a well-mixed digester will show different concentrations of contaminants in sludge samples taken from different elevations. Fats and grease tend to stratify toward the top of the digester along with most of the organics, cellulosic solids are spread throughout, and the mineral fraction that is not removed by the grit chambers settles to the bottom of the sludge column. Individual metals and organics behave differently depending on their molecular weights and composition: some tend to sink, some will adhere to solids or bind with other constituents, while others will float. Analyses of samples taken from the top and the bottom of a digester will reflect the full range of contaminant behavior.

2.3.2 Industrial Contributions

Contaminant concentrations are not constant, so any single sample of sludge is a function of time as well as digester elevation. Industrial contributions of contaminants are often introduced to the system as slug loadings from periodic inputs such as emptying a factory tank or washing contaminated surfaces at an industrial facility. It is commonly accepted that sludge levels of heavy metals and organics vary widely over time. Certain contaminants, such as mercury or cadmium, may be present in such small quantities that samples may vary in concentration by as much as three to fourfold and still be a reasonable reflection of actual sludge quality. Most of the perceived variation in metals concentrations is probably attributable to digester stratification and the vagaries of the sampling procedure, but industrial periodicity is occasionally cited as the culprit.

Slug loadings of industrial chemicals are most likely to be introduced to the wastewater system at the end of the week, at the end of the month, or on the weekends, when periodic tasks such as washing down floors or washing out tanks are most likely to be scheduled. To avoid industrial slug loadings in a sludge sample, the best bet is to sample in the middle of the week and the middle of the month. This way any slugs of contamination from the weekend or month's end periodic industrial loadings will have time to mix thoroughly with the rest of the digester.

2.3.3 Laboratory Error

Inaccurate results from laboratory testing of sludge also may contribute to perceived variations in sludge quality. Concentrations of metals and toxicants in municipal sludge are often as low as parts per million or even parts per billion, and laboratory results tend to be accurate only within a narrow range of values at these low concentrations.

Many sewerage districts plot their contaminant concentrations over a period of years to establish decreases in pollutants entering the system, using historical measurements of metals concentrations which may be misleading. Laboratories that have lowered their detection limits over years of sampling may appear to show reduced contaminant concentrations when they are actually reflecting improved laboratory techniques.

2.4 ANALYTICAL CHARACTERIZATION OF SLUDGE

Laboratory testing of sludge samples is one way to measure sludge contaminants, but it is not the only method available to a beneficial reuse program. Taking a representative sludge sample is not a straightforward exercise, and the laboratory results of sludge samples will not tell you if the contaminant concentrations in the sample are representative of the sludge quality in the digester. There are a variety of analytical tools available to confirm or refute concentration levels of sludge samples. Calculating the plant mass balances of contaminants and the theoretical removal rates of contaminants are two exercises which can be used to confirm that a sludge sample sent to the laboratory for testing is indeed a representative sample of the sludge in a digester.

2.4.1 Mass Balances of Contaminants

Mass balances are deceptively simple in concept: what goes in must come out. For a wastewater treatment plant, the influent loadings for contaminants must be equal to the contaminant loadings in the effluent and the sludge, providing there is no other significant waste stream (grit and screenings are not a significant pathway for most contaminants). This mass balance approach to characterizing sludge quality is the first step in confirming that the results of the sludge sampling program do in fact represent the sludge quality of the POTW.

The problem with doing a mass balance of contaminants through a treatment plant is that while the National Pollutant Discharge Elimination System (NPDES) permits require that POTW effluent be tested for priority pollutants, many treatment plants have sketchy or nonexistent data on influent flow. Since the influent flow is typically very well mixed, analyzing a few samples of the influent flow can yield a wealth of information. If the information on influent concentration levels is sparse, then an average annual concentration for any particular metal can be multiplied by the average annual flow. The chemical content of the influent flow is typically measured in concentrations (milligrams per liter) which must be converted into loadings per day (kilograms per day) by multiplying the influent concentrations by the average daily flow and converting. Using an influent loading of 3 mg copper per liter and an effluent loading of 2 mg copper per liter in a treatment plant with a flow of 1.2 MGD provides the information in Equations 2-1 and 2-2:

$$3 \text{ mg/l Cu} \times 1.2 \text{ MGD} \times 3.785 \text{ l/gal} = 13.6 \text{ kg/day Cu} \qquad \text{influent} \qquad (2\text{-}1)$$

$$2 \text{ mg/l Cu} \times 1.2 \text{ MGD} \times 3.785 \text{ l/gal} = 9.1 \text{ kg/day Cu} \qquad \text{effluent} \qquad (2\text{-}2)$$

Sludge loadings can then be calculated from this information (Equation 3):

$$\begin{aligned} \text{Sludge Cu (kg/day)} &= \text{influent Cu/kg/day)} - \text{effluent Cu (kg/day)} \\ 4.5 \text{ kg/day} &= 13.6 \text{ kg/day} - 9.1 \text{ kg/day} \qquad (2\text{-}3) \end{aligned}$$

Now we know that 4.5 kg of copper is entrained in the sludge per day. If this same treatment plant produces 8 dry tons of digested sludge a day, then the total copper retained in the plant can be divided by the total sludge production to get milligrams of copper per kilogram of sludge:

$$4.5 \text{ kg Cu} \div 8 \text{ dry tons sludge} \times 1{,}000{,}000 \text{ mg/kg} \times \text{ton}/2000 \text{ lb} \times 2.2 \text{ lb/kg} = 619 \text{ mg Cu/kg sludge} \quad (2\text{-}4)$$

If monthly measurements are made, plotting the metals loadings in the influent, effluent, and sludge on the same axis can be quite illuminating. While this exercise may seem painfully obvious, sludge that is sampled from the bottom of the digester may have metals loadings that are greater than the total influent loadings. This error in sampling can have disastrous effects on the options available for beneficial reuse, and it is easily corrected by improving the sampling program provided that measurements of the influent flows are made.

If the mass balance values do not roughly agree with the laboratory test results of contaminants in sludge samples (and particularly if the sludge samples are more heavily contaminated than is indicated by the mass balance calculations), further analysis is in order. Since both influent and effluent are considerably better mixed than the sludge, discrepancies in loadings would be a reasonable indicator that changes in the sludge sampling port or procedure could improve the sludge quality.

2.4.2 Theoretical Removal Rates of Contaminants

We now have two ways to determine sludge quality: by direct measurement and by projecting contaminant concentration from a mass balance of the influent and effluent flows. A third method of projecting sludge quality may be useful if there is little correlation between the values obtained from directly measuring the sludge quality and from calculating sludge loadings from the influent and effluent loadings.

In theory, primary and secondary wastewater treatment removes predictable levels of contaminants from the influent flow. Working backward from contaminant levels in the effluent and/or forward from influent levels, it is possible to project the expected levels of contaminants in the sludge with the information from Table 2-6. This method of checking sludge quality is quite rough, but should provide a ballpark figure to verify or discredit other estimates of contaminant concentrations.

For example, if the influent flow of 1.2 MGD to a plant carries 200 mg solids per liter and 3 mg copper per liter, the primary sludge production before digestion is theoretically

$$1.2 \text{ MGD} \times 8.34 \text{ lb/gal} \times \text{kg}/2.2 \text{ lb} \times 0.0002 \text{ solids} \times 0.5 \text{ solids removed} = 527.69 \text{ kg solids/day} \quad (2\text{-}5)$$

A theoretical 25% of the influent copper is removed, so the influent copper loadings of 13.6 kg/day × 0.25 copper removed = 3.4 kg/day. This copper loading is divided by sludge production to give Equation 2-6:

$$3.4 \text{ kg copper/day} \div 527.69 \text{ kg solids/day} = 644 \text{ mg/kg copper} \quad (2\text{-}6)$$

In this imaginary example, it is not surprising to see a close correlation between two analytical approaches to predict copper concentrations. In real life, numbers are rarely so tractable. If the laboratory results are far from the analytical values, the sludge samples are probably nonrepresentative. By using three methods to determine

Table 2-6 Wastewater Treatment Plant Removal Efficiencies

Pollutant	Primary treatment (% removed)[a]	Advanced primary treatment (% removed)[a]	Secondary treatment (% removed)
BOD	29–36	51	91
Solids	56–60	77	91
Antimony	30	NA	60
Arsenic	25	NA	50
Boron	2	NA	50
Cadmium	15	37	70
Chromium	27	69	80
Copper	25	64	80
Lead	50	NA	80
Mercury	25	57	80
Molybdenum	10	NA	50
Nickel	15	NA	35
Selenium	10	52	50
Silver	20	6	90
Zinc	30	62	80
Phenol	8	NA	95
Benzyl alcohol	NA	NA	90
1,2-Dichlorobenzene	NA	NA	90
2-Methylphenol	NA	NA	90
4-Methylphenol	NA	NA	90
Benzoic acid	NA	NA	90
Naphthalene	20	NA	95
2-Methylnaphthalene	NA	NA	95
2,4,5-Trichlorophenol	NA	NA	90
Dimethyl phthalate	24	NA	95
Diethyl phthalate	0	NA	95
N-Nitrosodiphenylamine	NA	NA	69
Di-*n*-butyl phthalate	20	NA	90
Butylbenzyl phthalate	50	NA	90
Bis(2-ethylhexyl) phthalate	0	NA	60
Di-*n*-octyl phthalate	0	NA	80
Fluorene	0	NA	90
Bromoethane	NA	NA	95
Methylene chloride	0	NA	50
Acetone	NA	NA	95
Carbon disulfide	NA	NA	95
trans-1,2-Dichloroethane	36	NA	90
Chloroform	14	NA	60
2-Butanone	NA	NA	90
1,1,1-Trichloroethane	40	NA	90
Trichloroethane	20	NA	90
Benzene	25	NA	90
4-Methyl-2-pentanone	NA	NA	90
Tetrachloroethane	4	NA	80
1,1,2,2-Tetrachloroethane	NA	NA	95
Toluene	0	NA	95
Chlorobenzene	NA	NA	90
Ethylbenzene	13	NA	90
Styrene	0	NA	90
Total xylene	NA	NA	95
Total PCBs	0	NA	92
Aldrin	0	NA	90
DDT	0	NA	90
Heptachlor	10	NA	90
Dieldrin	0	NA	90

Source: MWRA[6] and U.S. EPA.[7]

[a] NA = not applicable.

sludge quality — sampling and laboratory testing, mass balance calculations from influent and effluent flow concentrations, and the theoretical calculation based on predicted removal levels of specific metals — a sludge reuse program can verify contaminant concentrations or point out problems in the sludge sampling program.

REFERENCES

1. Metcalf & Eddy, Inc. *Wastewater Engineering: Treatment, Disposal and Reuse* (New York: McGraw-Hill, 1972).
2. Kuchenrither, R.D. and S.I. McMillan. "Preview Analysis of National Sludge Survey," *Biocycle* (July 1990).
3. 40 City Study — Sewage Sludge. "Fate of Priority Pollutants in Publicly Owned Treatment Works." National Sewage Sludge Survey, EPA 440/1-82-303. *Federal Register* 58(32): 9264, 9318 (1993).
4. Vesilind, P.A., G.C. Hartman and E.T. Skene. *Sludge Management and Disposal for the Practicing Engineer* (Chelsea, MI: Lewis Publishers, 1986).
5. "National Sewage Sludge Survey," U.S. Environmental Protection Agency (1988).
6. "Secondary Treatment Plant Facilities Plan," Massachusetts Water Resources Authority (1988).
7. "Supplemental Environmental Impact Statement of Wastewater Conveyance System for Boston Harbor," U.S. Environmental Protection Agency (1988).

FURTHER READING

Massachusetts Water Resources Authority, "Surveying the Nation's Sludge" *The Inside Scoop,* a Quarterly Sludge Update from the MWRA Residuals Management Team, Vol. 2 No. 1 (Spring 1991).

Massachusetts Water Resources Authority, "Treatment Technologies Fact Sheet, Boston Harbor Project," (1990).

U.S. Environmental Protection Agency, National Sewage Sludge Survey (1988).

U.S. Environmental Protection Agency, 40 City Survey (1978).

3 IMPROVING SLUDGE QUALITY

3.1 INTRODUCTION

The EPA 40 CFR Part 503 regulations provide limits on the concentrations of pollutants in sludge that is land applied. The most common hurdle to the beneficial reuse of sludge is the presence of unacceptably high levels of heavy metals. High metals concentrations in sludge that is land applied may limit the site life or reduce the application rate to the point where land application is uneconomical. When metals concentrations exceed the maximum regulatory limitations, land application is prohibited.

The EPA list of priority toxic water pollutants includes a wide range of organic and inorganic toxicants. Digestion and processing significantly reduces the number of coliform bacteria, bacterial pathogens, and parasites in sludge and the amount of volatile and semivolatile toxic organics, but it does not reduce levels of heavy metals. The industrial discharge of these substances to the sewer system is regulated by NPDES pretreatment requirements (40 CFR 403). These limitations and standards are established by regulation for categories of industrial dischargers and are based on the degree of control that can be achieved using various levels of pollution control technology. In addition, pretreatment standards are required for those pollutants which are not susceptible to treatment by POTW operation.

The national pretreatment program was established to regulate pollutants from nondomestic sources to local POTWs to improve effluent and sludge quality. The discharges targeted for regulation include those which interfere with the operation of a POTW (including interference with sludge digestion, reuse, or disposal), substances which will pass through the POTW, or substances which are incompatible with the treatment works. The pollution control strategy is based on three elements:

- National categorical standards: technology-based standards set by EPA headquarters which establish industry-specific effluent limits
- Prohibited discharge standards
  - General prohibitions (40 CFR 403.5[a]) — national prohibitions against pollutant discharges from nondomestic users which cause pass-through or interference with POTW operations
  - Specific prohibitions (40 CFR 403.5[b]) — national prohibitions against pollutant discharges from any nondomestic user causing (1) fire or explosion hazard, (2) corrosive structural damage, (3) interference due to flow obstruction, (4) interference due to flow rate or concentration, and (5) interference due to heat
- Local limits: Enforceable local requirements developed by POTWs to address federal standards as well as state and local regulations

The local limits, a critical tool for pretreatment programs, are specific requirements developed and enforced by individual POTWs that implement the general and specific prohibitions regulated in the prohibited discharge standards and go beyond them as necessary to meet state and local regulations. Local standards allow POTWs to meet the pretreatment objectives developed by those best placed to understand local concerns: the POTWs. In this scheme, the promulgation and enforcement of local limits is the critical link in ensuring that pretreatment standards protect both the local POTW and the local receiving environment.

The common approach to improving sludge quality is to strengthen industrial pretreatment programs, increase industrial monitoring, and tighten industrial discharge limits. Industry is not the only contributor of toxic materials to the sewer system, and a narrow focus on industrial contributions may not produce the desired results. Data from some treatment facilities suggest that the water supply system can be the major source of metals to a POTW. Other sources of contaminants to the influent flow of a wastewater treatment plant include households, urban runoff, and leachate from landfills or contaminated soils.

The origin of a toxic substance in the sludge must be determined before an effective strategy to reduce it can be formulated. For example, if sludge pesticide levels are high, the bulk of the contaminant may be coming from a manufacturer, the local municipal parks program, urban lawn care, or even agricultural runoff. Reduction strategies will be different for each source: reducing the industrial discharge limits will have little impact if the parks program is the primary source; a call to the parks commissioner will be ineffective if public education on lawn care is what is needed. By establishing a rough ranking of the relative contributions of each source for problem contaminants, significant improvements in sludge quality can be made at the lowest cost.

In theory, point source contaminants, or contaminants that are discharged to the sewer system at a single or few locations, are easier to control than non-point source contaminants. In practice, contaminant reduction is a multifaceted problem. Two examples of point sources of contaminants to the wastewater system are industries and academic laboratories. Industrial contributions are usually monitored by discharge permits and can be reduced through an industrial pretreatment program, but data from educational facilities often don't exist. Informing area schools of the effect that casual chemical disposal has on the local sludge reuse program is a strategy that may be effective, but if it is not there are few other plausible reduction strategies for that source.

Other sources of contaminants include the water supply, households, urban runoff, landfill leachate, and hazardous waste sites. Metals contributions from the water supply can be reduced by water treatment, providing a point source solution. Households, urban runoff, landfill leachate, and hazardous waste sites are sources of contaminants that are typically very difficult to reduce.

3.2 CALCULATING INFLUENT LOADINGS

The first step in sorting out sources of contaminants in the sludge is to quantify the influent loadings of toxics to the treatment plant. Most wastewater treatment plants have a better historical record for concentrations of influent heavy metals than for organics, since treatment plants have traditionally analyzed heavy metal concentrations along with the conventional wastewater treatment parameters such as biochemical oxygen demand, chemical oxygen demand, total suspended solids, etc.

Concentrations of contaminants in the influent flow are often so low that it is difficult to get accurate laboratory results, and detection limits may have to be refined. Laboratory analyses of toxic constituents are presented in Equation 3-1 as concentrations, which must be converted to daily loadings:

$$\text{Concentration (mg/l)} \times \text{flow (MGD)} \times 8.34 = \text{loading (lb/day)} \qquad (3\text{-}1)$$

In Boston, a city with little industry, lots of academic institutions, and acidic water, the collection system serves 2.2 million people and 43 communities. The total loadings of heavy metals to Boston's Deer Island Treatment Plant are shown in Table 3-1.

3.3 QUANTIFYING SOURCES OF CONTAMINANTS

Tracking down the sources of contaminants is a sleuthing process which takes a fair amount of ingenuity, and sometimes the process of elimination is the only available method of identifying a contaminant source. Contaminants attributable to industries and to the water supply can usually be roughly quantified, but it is very difficult to determine which fraction of the influent loadings is attributable to households and urban runoff (in cities with combined sewer systems) without a detailed data collection program.

If the levels of sludge contaminants are high enough to restrict options for beneficial reuse, a detailed sampling of the influent flow would be a useful tool in a contaminant reduction program. Sampling the flow at noon, evening, midnight, and morning is a method of determining which contaminants are background inputs and which are the result of human activity. If this detailed sampling is carried out over a period of time, it can also pinpoint which contaminants are sensitive to rainfall. In combined sewer systems, rainfall flushes contaminants from urban surfaces into the treatment plant and increases leachate contributions from landfills and hazardous waste sites. High flows will resuspend the sediment that settles to the bottom of the sewer pipes, further increasing contaminant loadings. The difference between dry and wet weather contaminant loadings can be attributed to urban runoff, leachate from landfills or hazardous waste sites, and flushing of the sewer lines.

Sampling at key points within the collection system is also an effective strategy to identify contaminant sources. Once specific areas within the collection system are identified as substantial contributors of contaminants, it is possible to narrow down the list of possible sources of contaminants. By narrowing down the geographic area of study, a sludge quality program can concentrate on a smaller set of possible problem contributors, optimizing resources.

3.3.1 Water Supply

As shown in Figure 3-1, the water system includes the reservoir which collects water and the municipal pipes which distribute it to the individual homes. Residential

Table 3-1 Influent Metals Loadings, Boston, Massachusetts

Metal	Influent loading (lb/day)
Chromium	45.8
Cadmium	15.3
Copper	287
Lead	80.2
Mercury	1.28[a]
Nickel	51.3
Zinc	442

[a] Below detection limits.

plumbing carries it from the street to the household taps. Metals and other contaminants are introduced into the water supply in two phases. As water leaches through the soil to fill the reservoir, metals are transferred from the soil to the water. Additional metals loadings are provided by the municipal pipes that deliver the water from the reservoir to the household, while the residential plumbing may be a significant source of copper and lead. Water treatment may or may not remove the metals from the water supply.

In the acid-rain belt of the U.S., the low pH of the water may leach significant amounts of heavy metals from both the soils and the pipes. Systems that are the most significant source of metals include those with

1. Low alkalinity — acidic surface water supplies
2. Low pH — high dissolved carbon dioxide groundwater supplies
3. Water systems that use zinc-phosphate corrosion inhibitors[1]

In areas where the industrial mix has few metal fabrication facilities, a substantial amount of the total loading of copper, zinc, cadmium, lead, and even mercury may be attributable to the water supply and the pipes. Sources of metals to the water supply are given in Table 3-2.

Contaminants attributable to the water supply usually can be quantified in part, since data on metals concentrations in reservoirs are usually collected annually or semiannually. Quantifying the fraction of contaminants that leaches from residential plumbing is a task requiring analyses of tap water samples from local homes. A less costly first step is to calculate the corrosivity of the water, which will indicate whether the residential plumbing is likely to be a significant source of contaminants.

The level of metals in the water supply is related to the corrosivity of the water. Corrosivity is a nonintuitive quality that is measured by a number of indices including the aggressive index, or AI, and the Langlier saturation index, or LSI. Both of these indices estimate the tendency of a water to lay down a protective coating of calcium carbonate ($CaCO_3$) on the pipe wall. A thin layer of $CaCO_3$ keeps the water from contacting the pipe, which is the only practical way to prevent corrosion. The $CaCO_3$ saturation value of the water depends on the calcium ion concentration, alkalinity, temperature, pH, and the presence of other dissolved minerals such as chlorides and sulfates.

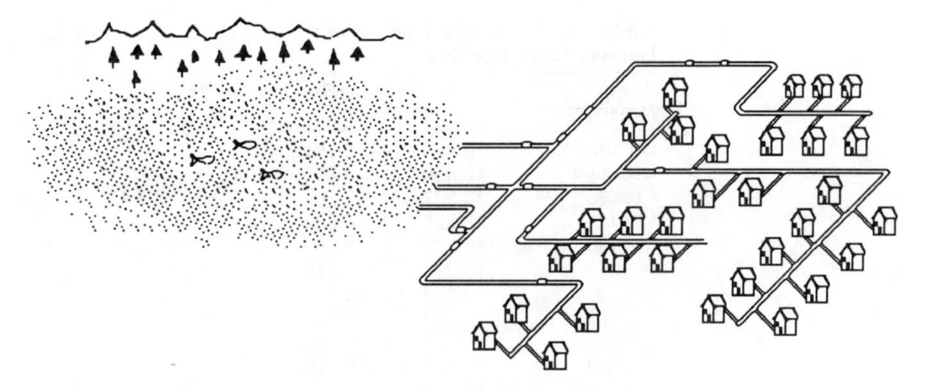

Figure 3-1 Municipal water supply system.

Table 3-2 Sources of Metals in the Water Supply System

Copper
Copper house services
Copper plumbing
Zinc
Galvanized piping
Plated fixtures
Zinc phosphate corrosion inhibitors
Cadmium
Galvanized pipe (cadmium is a common impurity in zinc)
Lead
Lead service pipes
Lead gooseneck connections on water mains
Lead/tin solder used to join copper piping.
Mercury
Solder joining copper pipes

Source: Brown.[3]

3.3.1.1 Langlier Saturation Index

The LSI is a widely used index to determine the corrosivity of the water.[2] Based on the effect of pH on the solubility of calcium carbonate, it provides a measurement that predicts whether a protective scale layer will be deposited or dissolved by the water. The information necessary to use the LSI includes total alkalinity, calcium (milligrams per liter), total dissolved solids, pH, and temperature. These factors are combined in Equation 3-2 to determine the value of the LSI:

$$\text{LSI} = \text{pH} - \text{A} + \text{B} - \log[\text{Ca}^{++}] - \log[\text{total alkalinity}] \quad (3\text{-}2)$$

where A and B are constants related to the temperature and the dissolved solids of the water, found in Table 3-3. The values of the LSI are interpreted as follows:

LSI > 0 Noncorrosive; the water will lay down a protective coating of $CaCO_3$, protecting the pipes from corrosion.

Table 3-3 Constants A and B of the Langlier Saturation Index

Temp (°C)	A	B
0	10.627	8.022
10	10.490	8.151
18	10.400	8.260
20	10.377	8.280
25	10.329	8.342
30	10.290	8.395
40	10.220	8.514
50	10.172	8.625
60	10.143	8.738
70	10.124	8.860
80	10.112	8.975
100	10.070	—

LSI = 0 Neutral; there will be no protective layer deposited on the pipes, but neither will the water corrode the pipes.

LSI < 0 Corrosive; the water will tend to dissolve all $CaCO_3$, and the pipes will corrode.

3.3.1.2 Aggressive Index

The AI was developed to help engineers select the appropriate pipe material to ensure the long-term structural stability of the system.[1] The AI is a simplified form of the LSI; it approximates the solubility of calcium carbonate rather than the corrosivity of the water. This index uses pH, total alkalinity (A), and calcium hardness (H) in Equation 3-3:

$$AI = pH + \log[A/H] \tag{3-3}$$

The values are interpreted as follows:

AI < 10 Very aggressive
AI = 10–12 Moderately aggressive
AI > 12 Nonaggressive

Either the LSI or AI should be used to provide a reading on whether pipe corrosion is likely to be a significant source of metals to the treatment plant. If the water is corrosive, the household tap water within the collection area should be tested for concentrations of heavy metals to establish the contribution of the pipes and plumbing to the treatment plant.

Metals concentrations in the water supply are usually measured annually to comply with laws regulating drinking water quality. Starting at the reservoir, these metals concentrations should be multiplied by the annual volume of water supplied and converted to daily loadings to provide estimates for heavy metals attributable to the reservoir.

The metals that leach from the pipes — the so-called tap water metals — are more difficult to quantify than the metals attributable to the reservoir and will likely require special data collection. Not only are data on tap water quality often nonexistent; there are significant problems in collecting meaningful measurements: the first flush effect is profound with corrosive tap water, and the metals concentrations in water left standing overnight in the pipes may be an order of magnitude higher than that of running water from the same tap.

Shown in Table 3-4 are the data for reservoir and tap water loadings in Boston. The tap water samples were taken on three days in March 1988. A total of 87 samplers collected two samples each: one during the first few seconds that the water was running, and one after the faucet had been on for 5 min. The variability in the first flush samples was so great that the values in the table were derived only from running water.

As is shown in this example, the contribution of metals from the water supply and pipes can be surprisingly high. One quarter to one half of the total loadings of some of the metals can be attributed to the water supply.

3.3.1.3 Reducing the Metals Attributable to the Water Supply

The Safe Water Drinking Act of 1974 called on the EPA to determine which substances were contaminating the nation's water supply, to assess the potential effects, to establish legal health limits, and to intervene where the states fail to enforce them. There are a total of 56 standards in place, and utilities must test their water to determine compliance. If analyses of water corrosivity reveal that the water is corrosive, there are well-established treatment practices that can be implemented by the water supply system to reduce the levels of metals. The major obstacle to reducing the concentrations of metals attributable to the water supply is institutional rather than technical. Most water supply systems and wastewater facilities are controlled locally by separate boards or districts and are regulated by separate agencies or groups on the state and federal levels. In many instances it is difficult to convince the local water authority to incur an expense to benefit the local wastewater treatment facility, but a little interagency cooperation can have a big impact on sludge quality. A small treatment plant in Ellsworth, ME ran analyses of their sludge before and after adjusting the pH of their water supply, and the results, found in Table 3-5, speak for themself.

It is evident that sludge lead levels seem to be minimally affected by the pH adjustment of the water, while copper, zinc, and cadmium levels in sludge are substantially reduced. Zinc concentrations in sludge were reduced by 35% when the pH was raised from 6.3 to 7, but further adjustments of the pH had little impact. Copper and cadmium concentrations in sludge fell by 25 and 30%, respectively, when the pH was raised from 6.3 to 7.0 and fell again when the pH was raised to 8.0. The total reductions in sludge copper and cadmium concentrations by raising the pH from 6.3 to 8.0 were 60 and 55%, respectively.[3] A 74% decrease in copper concentration in the drinking water in Fort Collins, Co resulted in a 25% reduction in sludge copper concentrations.[4]

3.3.2 Industrial Contributions

Industrial contributors to the wastewater system are required to meet all applicable national pretreatment standards, which may include general and specific prohibited

Table 3-4 Metals Loadings Attributable to Boston Water Supply

Metal	Reservoir (lb/day)	Tap water (lb/day)	Total influent (lb/day)
Chromium	12.1[a]	NA	45.8
Cadmium	NA	2..3	15.3
Copper	60	127	287
Lead	4.8[a]	13.0	80.2
Mercury	0.5[a]	NA	1.28[a]
Nickel	12	NA	51.3
Zinc	96.8	NA	442

Source: MWRA.[8]
Note: NA = not applicable.
[a] Below detection limit.

Table 3-5 Sludge Metal Levels Before and After pH Control of the Water Supply in Ellsworth, ME

	Date			
Metal	10/27/85 pH = 6.3	9/24/87 pH = 6.3	5/14/86 pH = 7.0	7/17/86 pH = 8.0
Cu	978	902	737	378
Zn	806	1048	522	528
Cd	10.5	9.0	7.5	4.6
Pb	324	—	311	296

Source: Brown.[3]
Note: Values are in parts per million.

discharges standards, categorical pretreatment standards, and local limits. In addition, POTWs designed to accommodate design flows of more than 5 MGD and smaller POTWs with significant industrial discharges are required to establish local pretreatment programs. The local program must include adequate legal authorities, industrial use permitting, compliance monitoring, enforcement, and public participation. Every industrial discharger is required by law to comply with the pretreatment standards of its district, although compliance may depend on the level of oversight provided by the wastewater district.

Industrial contributions can be a well-documented source of contaminants to the wastewater. Data provided by the discharge permits provide a baseline of contaminant inputs to the system. If the existing pretreatment program is sketchy or nonexistent, substantial improvements in influent quality may be attained in a short period of time by focusing attention on this source of contaminants.

3.3.2.1 Sewer Use Limitations

The Clean Water Act of 1972 requires the institution of pretreatment programs to control the discharges of industrial waste, and many POTWs are beginning to exercise control over the toxic constituents entering their system via the NPDES pretreatment requirements of 40 CFR 403. The pretreatment requirements are based on a three-part strategy. First, the national categorical standards provide nationally uniform effluent limits affording a technology-based degree of environmental protection for discharges from particular industrial categories. Second, the prohibited discharge standards recognize the site-specific nature of POTW concerns and provide a broader baseline level of POTW control, regardless of whether the industry falls within a specific category. Third, local limits are specific requirements developed and enforced by individual POTWs implementing the general and specific prohibitions and going beyond them as necessary.

Pretreatment programs, as defined by 40 CFR 403, are typically based on end-of-the-pipe concentration limits for the various constituents present in industrial discharges. This approach is consistent with water quality concerns, but it is not consistent with the goals of a beneficial sludge reuse program. Efforts to improve water quality generally involve reducing the toxics in the effluent, and the conventional criteria used to measure effluent quality are the concentrations of contaminants (milligrams per liter). Efforts to improve sludge quality involve reducing the toxic constituents entrained in the sludge, a significantly different problem. A properly operating treatment plant will transfer most of the pollutants out of the effluent stream and into the sludge, so the conventional criterion to improve sludge quality is the loading rate of pollutants (pounds per day) to the system. This means that the controls used to improve effluent quality will not necessarily improve sludge quality, since an industrial facility need only increase the water flow to reduce the concentration of contaminants, even though total loadings can be raised.

When a POTW establishes local limits for the concentration of regulated chemicals to limit industrial discharges of pollutants to the collection system, it provides a method of controlling water quality. However, concentration-based limits do not address the issue of pollutant *loadings* to the treatment facility. For a POTW to set concentration limits that will improve sludge quality, the concentration limits must be based on total loading to the treatment plant rather than on effluent concentrations.[5]

3.3.2.2 Industrial Loadings

Records of industrial discharges of toxic constituents are generally available. However, there may be several thousand businesses that must be included in the analysis. Sorting out industrial contributions of contaminants is a time-consuming process, and only constituents at levels which are barriers to beneficial reuse deserve this detailed analysis.

3.3.3 Households

A wide range of commercial household products are designed to be disposed of down the drain, including detergents, household cleansers, disinfectants, shampoo, soap, and drain openers. Household hazardous waste such as pesticides, paint products, and automotive products also find their way to the treatment plant through thoughtless or irresponsible disposal.

While each individual may use only small amounts of these substances, the cumulative loadings of household products could be a substantial source of contaminants to the treatment plant. At the same time, the pollutants are from diffuse sources, making them very difficult to control. A single household might wash less than 0.5 lb of heavy metals down the drain a year, but the treatment plant in a town of 50,000 would annually process an additional 900 lb of heavy metals, while the wastewater of a city of 2 million people would have to process 18 tons of these toxins a year, most of which end up in the sludge. The contributions of hazardous household products of a single person to the wastewater are small, but the aggregate loadings of a community may be substantial.

Household cleaners often contain toxic or hazardous substances. Organic solvent constituents appear to be the most toxic, especially disinfectants such as chlorinated phenolics (little is known about their long-term effects) and dry cleaning or septic tank/drainfield degreasing solvents such as 1,1,1-trichlorethane, which is known to produce a wide range of chronic effects and has contaminated groundwater in some parts of the country. Detergents are used in enormous quantities and may be significant sources of nickel to the treatment plant. Disinfectants designed to kill microorganisms contain pesticide-type components. Solvents such as nail polish removers also may find their way down the drain.

Paint products as a class appear to pose relatively little potential for environmental harm. They are not usually dumped down the drain, and paints usually don't contain chemicals with long-term effects. Many solvents are toxic to aquatic organisms in low concentrations and may bioaccumulate. The persistent chlorinated hydrocarbons are particularly noxious. Wood preservatives such as pentachlorophenol, while generally considered to be paint products, are extremely toxic pesticides and are usually resistant to breakdown under natural conditions.

Significant quantities of motor oil are used in each community, and the contributions of waste oil provided by weekend mechanics pouring used oil down the gutter may be significant in areas with combined sewer systems, although these are not as common as they were in earlier years. Gas stations with improperly operating oil/gas separators also can contribute significant amounts of waste oil to the treatment plant. Petroleum products are hazardous environmental pollutants, with problems associated with their physical characteristics (coating surfaces, flammability) as well as toxicological effects (both acute lethal and long-term sublethal). Waste petroleum products also can accumulate in sediments and bioaccumulate in organisms. Ethylene glycol in antifreeze, another ubiquitous chemical, is acutely toxic in a concentrated form, but appears to be adequately broken down if diluted in the sewage system or soil.

Pesticides are also dumped down the drain, and they come in a wide variety of classes. The persistent chlorinated hydrocarbon pesticides such as DDT have given way to the more quickly degraded and more acutely toxic carbamates, organophosphates, and natural product extracts. All are dangerous to aquatic life and environmental quality.

Table 3-6 provides a rough estimate of the per capita concentrations of a limited number of household contaminants in the wastewater system. Multiplying the per capita contribution by the population in the collection area provides a very rough estimate of the fraction of heavy metals attributable to household activities. In the Boston area, wastewater from 2.2 million people flows into the wastewater treatment plant, Table 3-7 provides a rough estimate of some of the contaminants attributable to households in that area.

Household contributions of toxics to the wastewater can be reduced by clean sweep programs, education, or by-product reformulation — by one uphill battle or

Table 3-6 Household Product Contributions of Heavy Metals

Product	Household contributions (μg per capita per day)						
	Cd	Cr	Cu	Pb	Hg	Ni	Zn
Mouthwash	0.3	2.1	1.7	1.2	0	69	2.9
Toothpaste	4.2	11.3	26.4	29.3	0.1	326	17.3
Toilet paper	4.5	21.9	3.8	23.9	0.6	108	31.3
Bath soap	18.3	267	104.7	141.2	6.3	—	117.9
Shampoo	0.2	4.0	1.3	0.6	0	32	44.9
Shaving cream	0.1	1.8	0.7	0.2	0.1	18	1.2
Liquid dish detergent	4.5	22.2	32.8	2.0	0	536.6	5.6
Dishwasher detergent	1.3	13.1	7.3	2.8	0	—	12.8
Kitchen cleanser	0.6	11.9	13.4	3.0	0	—	312
Toilet bowl cleaner	0.3	6.3	4.5	0.9	0	283.9	30.8
Powder laundry soap	26.6	221.9	249.4	31.3	0.3	—	94.5
Liquid bleach	3.6	10.7	5	2.0	0	519.6	8.1
Powder bleach	2.5	22.7	24.3	1.5	0	—	11.2
Liquid household cleaner	1.7	7.6	3.4	2.1	0	92.4	10
Totals	**68.7**	**624.5**	**478.7**	**242**	**7.4**	**1985.5**	**700.5**

Source: U.S. EPA.[9]

the other. Industry, not Mom, is perceived to be the source of water pollution. In Vermont, grocery stores post a poison label on racks holding hazardous household products to alert consumers of the impact of consumer choice on the environment. Increased awareness of the impacts of hazardous substances and the formulation of more "green" products should ultimately result in a reduction of household contributions of contaminants to the treatment plant.

3.3.4 Runoff

The fraction of contaminants attributable to runoff is particularly difficult to quantify. In a city with separate storm and sanitary sewers, the influent flow to the POTW should not be significantly impacted by runoff. In cities with combined sewer systems, runoff can be a significant source of contaminants to the treatment plant. Contaminants associated with runoff include car exhaust products and other airborne particulates that settle on the roadways, tire dust that collects on the asphalt, oily residue from automobiles, pesticides and fertilizers from construction sites, lawns, and gardens, asbestos from brakes, clutches, and tires, ethylene glycol, animal feces, etc. These contaminants accumulate on paved surfaces and do not enter the collection system until they are washed into combined sewers by rainfall or street washing.

Samples of influent flows under wet weather and dry weather conditions show a marked difference, with the wet weather samples containing significantly higher levels of most contaminants. The difference between these loadings can be attributed to runoff.

The U.S. EPA Storm and Combined Sewer Research Program found that runoff samples varied widely in their relative toxicities. A residential roof runoff sample was found to be the most toxic, attributed to metals leaching from galvanized roof

Table 3-7 Household Contributions of Metals, Boston, Massachusetts

Metal	Contribution (lb/day)
Cadmium	0.33
Chromium	3.02
Copper	2.3
Lead	1.17
Mercury	0.036
Nickel	9.61
Zinc	3.39

gutters and downspouts and to pesticide applications that were rinsed from the roof. Other samples with relatively high toxicities were from automobile service facilities, unpaved industrial parking and storage areas, and paved industrial streets. The most common toxicants include heavy metals, pyrene, fluoranthene, and 1,3-dichlorobenzene.[6]

Toxicants washed from urban surfaces are one source of contaminants associated with wet weather conditions. The other source is the contaminated sediment that settles to the bottom of the sewer pipes during low flow conditions and is swept up with the influent flow during a storm event, the so-called first flush effect. Wet weather flows can be 10 times greater than dry weather flows and, all things being equal, influent contaminant concentrations during high flow conditions would be lower than the contaminant concentrations during low flow conditions. Instead, monitored mass loadings show that significant quantities of toxics are flushed into the system during an urban rainstorm, and both contaminant concentrations and total suspended solids spike to well above dry weather levels. Regular sewer cleaning and landfilling of the contaminated sediments dredged from the sewer pipes is a method of reducing the contaminants associated with the first flush effect, although this does not affect the runoff contribution.

Urban runoff occurs for relatively short periods of time. The loadings are periodic, and an individual rainstorm event provides a pulse input of contaminants that changes rapidly over time. In the first few minutes of a rainstorm, the contaminants on the roadway are swept into the drains in a first flush which can have a pH as low as 2. Concentrations fall off rapidly after the first flush has cleaned the paved areas that drain into the collection system and rinsed accumulated solids from the sewers. This variation in contaminant concentrations and levels of total suspended solids, coupled with a surge of influent flow associated with a rainstorm event, makes it very difficult to construct an adequate sampling program to directly measure the contributions of runoff. Careful monitoring of the influent flow yields more useful data than sampling runoff at the source.

Combined sewer overflows, where mixed storm and sanitary flows are shunted to the waterways during storm events, are slowly being phased out. Cities that undertake structural changes to their collection systems to correct the water quality problems created by the introduction of untreated sewage and street runoff to their water bodies usually shunt that overflow to the treatment plant. The inclusion of runoff to the influent flow of the treatment plant can increase contaminant loadings to the POTW and may adversely affect sludge quality.

3.3.5 Leachate from Landfills and Hazardous Waste Sites

Leachate from landfills and hazardous waste sites may contribute to the contaminant loading at the wastewater treatment plant. Leachate may be directly discharged to a treatment plant or may be flushed into the collection system during a rainstorm event.

3.4 CASE STUDY — SOURCES OF PETROLEUM HYDROCARBONS

The contamination problems due to petroleum hydrocarbons (PHCs) have long been recognized in the wastewater treatment industry. Both surficial and underground oil spills can be carried to treatment facilities and affect effluent, sludge, and air quality at the treatment plants. Identifying the sources of PHCs to the treatment plant can be a difficult and time-consuming task. PHCs are commonly used materials and are found throughout a collection area. They enter the collection system by industrial contributions, residential discharge, urban runoff, and infiltration. However, they most probably do not enter the system from the water supply or residential plumbing.

PHCs that enter the collection system can impact the quality of sludge and effluent. Although petroleum hydrocarbons are usually considered to be a problem for effluent toxicity, daily variation studies have shown that PHCs in the form of fuel oils, kerosene, and waste oil are routinely present in sludge in relatively high concentrations in most industrialized service areas. Most of the time, volatile compounds are transferred to the air during collection and treatment process, but some volatile and many semivolatile compounds are detectable in the sludge. If the sludge is to be beneficially reused, high levels of PHCs may reduce the available options for beneficial reuse.

3.4.1 Calculate Influent Loadings of Petroleum Hydrocarbons

In May of 1989, a study was conducted to locate and identify the sources of PHCs in the Massachusetts Water Resources Authority (MWRA) service area in an effort to improve sludge quality. The sampling stations were chosen based on the ability to provide information on separate regions of the collection system. Four headworks that served different areas were tested, and the influent flow at the treatment plant was sampled. The samples were analyzed to identify and quantify the PHCs entering the collection system. Based on the overall network for the collection system, the following flow balance can be written:

Winthrop +	Chelsea +	Columbus +	Ward Street =	Deer Island
Terminal	Headworks	Headworks	Headworks	Treatment Plant

Although there are some additional connections to the collection system, this flow balance represents over 90% of the flow to the treatment plant. To account for variations in daily activities, a programmed automatic sampling device was used to collect samples every 6 h over a 24-h period for 10 days throughout the month of May in 1989.

The PHCs that enter the system through infiltration and underground oil plumes provide a background loading of oil which is relatively constant. PHC inputs that vary diurnally include industrial discharges, residential discharges, floor washing from service stations, overflowing oil/water separators, and commercial laundries. Finally, the PHCs attributable to urban runoff correspond to wet weather conditions.

Table 3-8 presents the dates corresponding to weekends and rainy weather conditions during the 1-month study. Figure 3-2 presents the PHC loading to the Deer Island Treatment Plant in relation to time of day. During the evening hours, the PHC loading to the treatment plant is significantly higher. The sharp peaks on May 4 and May 10 correspond to rainy weather conditions.

3.4.2 Quantify the Sources of Petroleum Hydrocarbons

It can be assumed that the water supply, residential plumbing, and academic laboratories are not a significant source of PHCs. Permitted industrial contributions can be quantified easily: in this case, the contributors of PHCs to the treatment plant include a total of 54 industries, most of which are industrial laundries and machined parts washing and electroplating operations. The permitted industrial discharge of PHCs to the wastewater system was found to be 944 lb/day. This leaves the contributions from runoff, households, and leachate to be determined.

Figure 3-3 presents the PHC loadings to the headworks and the Deer Island Treatment Plant as a function of sampling dates and the time of day. From Figure 3-2 it is evident that Ward Street contributes the largest fraction of PHCs to the Deer Island Treatment Plant, averaging 660 to 2220 lb/day from morning to midnight, with the early morning loadings substantially less than those later in the day. Chelsea headworks contributed the second largest amount with a PHC loading of 600 to 2100 lb/day. Analysis showed Winthrop terminal to be only a minor contributor of PHCs to the system. Columbus Park, with about half the flow and half the loadings of Ward Street and Chelsea headworks, also showed a distinct difference between morning and evening loadings, which ranged from 350 to 1435 lb/day. The increase in PHC loading during evening hours can be traced to the end-of-the-day cleaning operations by industrial services.

PHC loadings to the collection system varied depending not only on the time of day, but on the weather conditions as well. The two peaks corresponding to the rainy days at the Deer Island influent show the significance of PHC contamination due to runoff. Table 3-9 summarizes the variation in PHC loading to Deer Island Treatment Plant in relation to weather conditions. Figure 3-4 presents this same data in graphical format.

As shown in Figure 3-5, there is a significant difference between the weekday and weekend patterns. During dry weather conditions, the PHC loading peaks were observed during evening hours. However, during weekend conditions, the PHC loadings were higher during evening and midnight hours.

The data reveal that the major sources of PHCs to the collection system are runoff and industrial sources, as shown in Table 3-10. The loadings of PHCs vary with weather conditions, time of day, and day of the week. The difference between the morning and evening loadings represents oil from industrial or household activities. We can see that the average daily loadings at the Deer Island Treatment Plant range

Table 3-8 Sampling Dates on Weekends or During Rainy Weather

Date	Rainy weather	Weekend
5/1		
5/4	X	
5/7		X
5/10	X	
5/13		X
5/16		
5/19		
5/22		
5/25		

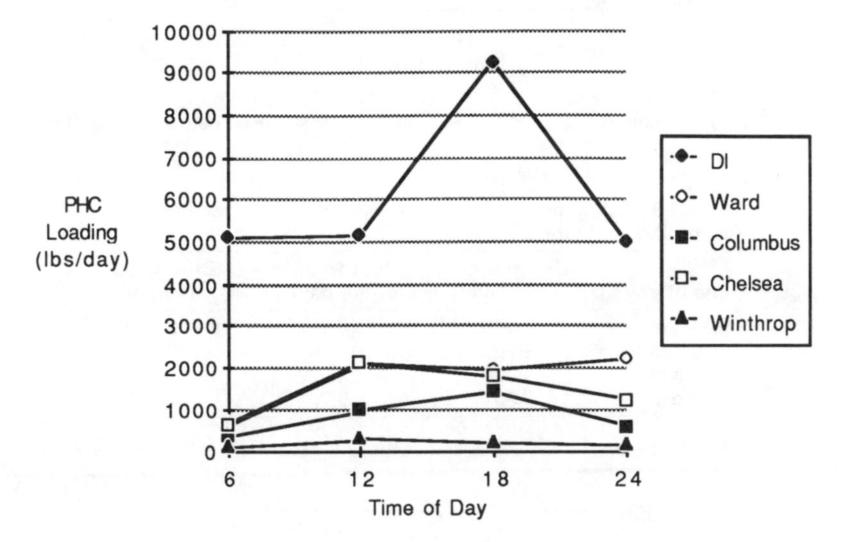

Figure 3-2 Petroleum hydrocarbon (PHC) loadings in relation to time of day.

from over 4000 lb/day in the morning to nearly 9000 lb/day in the evening. The loadings of PHCs are much higher in the evening than in the morning, indicating that oil enters the collection system as the result of human activities.

In dry weather, the 6 a.m. loading of PHCs totals 2400 lb/day. This could represent the background loading attributable to passive sources such as leaking underground storage tanks and other oil spills.

PHC loadings are much higher during wet weather, indicating that runoff is a major contributor. PHCs enter the collection system later in the day during weekends than they do during weekdays, which may point to the activities of "sidewalk mechanics", or householders that change their motor oil and dispose of it in the collection system.

Unpermitted industrial sources such as overflowing oil/water separators are a likely source of PHCs. There are 2000 oil separators used throughout the service area by parking garages, gas stations, car washes, radiator shops, car dealerships, and truck terminals. These oil separators are designed to capture a mixture of oil and water from the floor drain, retaining the hydrocarbons while draining the water to the sewer through the bottom of the tank. They are designed to be used only as precautionary devices, but regular inspection has shown that the gas oil separators are also

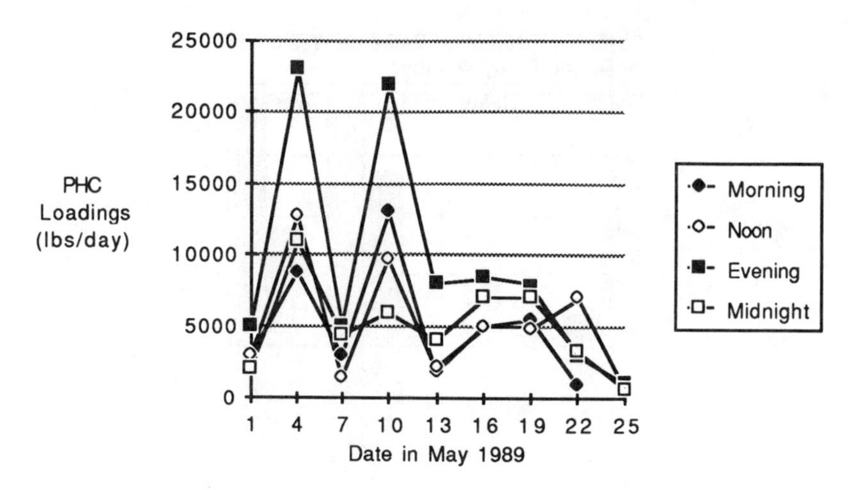

Figure 3-3 Petroleum hydrocarbon (PHC) loadings by sampling date and time of day.

Table 3-9 Difference in Petroleum Hydrocarbon Loading During Dry and Wet Weather

Time of day	Dry weather (lb/day)	Wet weather (lb/day)	Difference (lb/day)
6:00 a.m.	2,400	10,800	8,400
12:00 a.m.	3,700	11,400	7,700
6:00 p.m.	4,900	22,000	17,100
12:00 p.m.	3,900	8,700	4,800

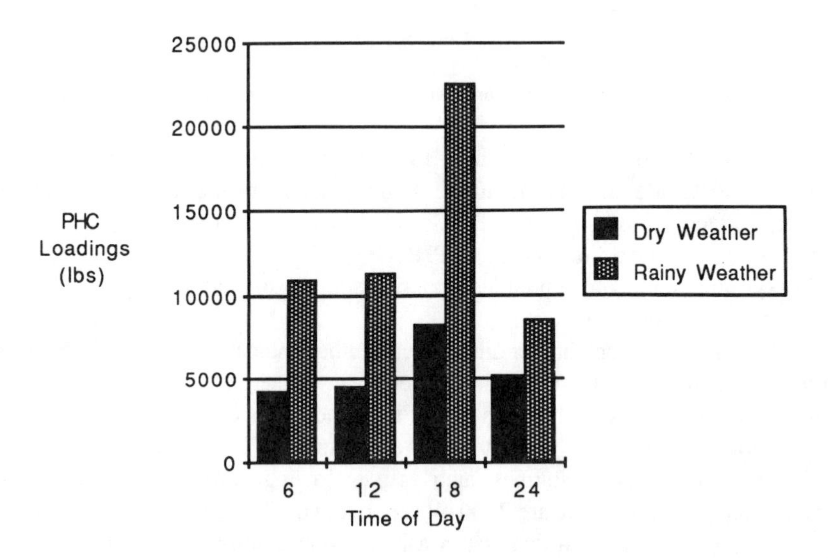

Figure 3-4 Effect of runoff on petroleum hydrocarbon (PHC) loadings.

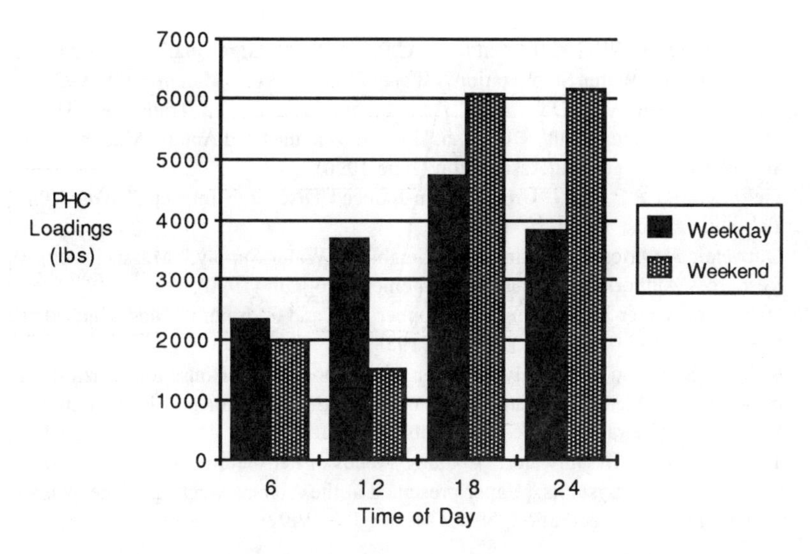

Figure 3-5 Effect of weekday or weekend activities on petroleum hydrocarbon (PHC) loadings.

Table 3-10 Sources of Petroleum Hydrocarbons to the Deer Island Treatment Plant

Source	PHC loading (lb/day)
Industrial sources[a]	944
Infiltration/inflow	600
Unaccounted	2,500
Urban runoff[b]	9,500
Total (dry weather)	4,000
Total (wet weather)	13,500

[a] PHC loading attributed to permitted industrial discharges.
[b] Wet weather only.

being used to dispose of waste oil, lubricating oil, spent automotive coolant, and discharges from steam cleaning operations. Should each of these devices be releasing as little as a cup of oil a day to the system, it would be a significant source of PHCs to the system.

REFERENCES

1. Degrément, G. *Water Treatment Handbook* (Caxton Hill, Hertford, England: Stephen Austin and Sons Ltd., 1973).
2. Tchobanoglous, G. and E.D. Schroeder. *Water Quality: Characteristics Modelling Modification* (Reading, MA: Addison-Wesley Publishing Company, 1985).
3. Brown, W.E. "Water Supply and Sludge Metals," *Biocycle* (April 1988).

4. Kuchenrither, R.D., G.K. Elmund, and C.P. Houck. "Sludge Quality Benefits Realized from Drinking Water Stabilization," *Water Environ. Res.* (March/April 1992).
5. Tansel, B. and A.B. Outwater. "Pretreatment Programs in Transition: The Water Quality vs. Sludge Quality Dilemma," Presented at the 83rd Annual Meeting of the Air and Waste Management Association (June 1990).
6. Field, R. and R.E. Pitt. "Urban Storm-Induced Discharge Impacts," *Water Environ. Technol.* (August 1990).
7. Outwater, A. "Metals Loadings Attributable to Water Supply," Massachusetts Water Resources Authority, Residuals Management Program (1990).
8. "Report on Water Testing for Lead, Copper, Iron and Cadmium," Massachusetts Water Resources Authority Water Division (1988).
9. Ridgley, S.M. and D.V. Galvin. "Summary Report of the Household Hazardous Disposal Project, Metro Toxicant Program #1", U.S. Environmental Protection Agency, Washington Operating Office (December 1982).
10. Tansel, B. and A.B. Outwater. "Loading Trends of Petroleum Hydrocarbons to Wastewater Collection Systems," Paper presented at the Annual Meeting of the Wastewater Pollution Control Federation, New Orleans, LA (1992).

FURTHER READING

Egan, J.T. "Toxics Control: The Municipal Perspective," *Water Environment and Technology* (March 1991).

Massachusetts Water Resources Authority Residuals Management Program, "Evolution of the Deer and Nut Island Sludge Monitoring Program" January, 1990.

Schultz, W. and P. Yarossi. "The Reduction of Copper and Oil in the Massachusetts Water Resources Authority Treatment Plant Sludges," Massachusetts Water Resources Authority, Toxics Reduction and Control (1990).

Tighe and Bond, Inc. *Interim Sludge Processing and Disposal Project, Final Facilities Plan, Environmental Impact Report,* Vol. II, Prepared for the Massachusetts Water Resources Authority (April 1989).

4 MINOR RESIDUALS

4.1 INTRODUCTION

The minor residuals of wastewater treatment include scum, grit, and screenings. Grit is invariably landfilled, and there is virtually no reuse of this material. Screenings are usually landfilled and occasionally ground and released to the digester, a management practice that is falling out of favor since screenings digest very poorly. Like grit, screenings are not reused. Scum, on the other hand, is largely biodegradable, and coprocessing the scum with sludge is the preferred method of managing this residual. Scum is the source of much of the methane gas produced during digestion, so scum use has come to mean collection of the methane produced during digestion, a standard feature at many wastewater treatment plants. Improved digester mixing and improved pretreatment serve to minimize scum production and maximize gas production.

There are unique handling problems associated with scum. The floating grease coagulates at room temperature and will routinely clog pipes, pumps, and digesters. To eliminate clogging and adhesion problems, heated pipes are often installed when scum is processed. Scum has a particularly noxious odor, mingling the scent of rancid grease with an overtone of raw sewage. The repulsive nature of scum makes it a discrete disposal problem for all plants, regardless of size.

4.2 SCUM CHARACTERIZATION

Scum floats on the surface of the wastewater and could be considered to be the aggregate bathtub ring of the wastewater collection area. Composed of varying quantities of skin, soap, grease, vegetable and mineral oils, bits of wood, paper, and cotton, and a variety of "personal hygiene items" — including adhesive bandages, plastic tampon applicators, and condoms —scum is a particularly unappealing residual.[1]

Scum is notoriously difficult to sample. Made up of grease and entrained solids that are suspended in a wastewater matrix, a scum sample will always stratify in wastewater. Unthickened scum collected during primary processing may consist of from 2 to 10% solids, depending on whether the sample is taken from the bottom or the top of the collection tank. Thickened scum may vary from 5 to 65% solids, posing even more difficult sampling problems. The greatest variable in scum solids level determination may be the laboratory procedure for processing scum samples. Laboratory personnel should process the largest possible scum sample to determine solids content, for there is no simple method to partition bottles of stratifying scum accurately.

Regularly scheduled pump-outs from the storage digesters are also used to estimate scum production figures. This measures the digested scum and invariably includes more material than the plastics that remain after anaerobic digestion. If scum production figures are derived from simple draw-off from a thickening tank, there may or may not be conscious attempts to minimize excess water. Finally, scum production figures provided by various treatment plants do not differentiate between raw and digested scum.

4.2.1 Physical Composition

Table 4-1 provides scum production data from treatment plants around the country. The quantities vary widely, as should be expected considering all of the factors confounding scum data collection.

It is evident from the ranges in scum values that the sampling procedures used by these treatment plants are probably wildly divergent. The volumetric and dry solids mass per million gallons of plant flow varies widely, while there is some narrowing (and a much smaller data base) for the volatile solids content and the BTU per pound solids.

4.2.2 Chemical Composition

Scum is typically coprocessed with the sludge, so the procedures for scum sampling and sample handling are less standardized than they are for sludge. As a result, it is difficult to make meaningful comparisons between scum analyses from different POTWs. Table 4-2 presents figures for scum quality from samples taken from the same scum thickening tank 5 years apart.

4.2.3 Flotables

Scum contains flotables, which include most of the plastics that reach the wastewater treatment plant. Plastics articles in the influent flow include plastic tampon applicators, condoms, and adhesive bandages. Of these, only plastic tampon applicators have the structural integrity to survive the journey through the wastewater treatment plant intact and to persist in a recognizable form in the environment. These applicators are one of the most intractable problems in residuals management. Plastic tampon applicators are the most visible fraction of wastewater residuals and are so common on beaches along the East Coast that they are referred to as "beach whistles". They degrade very slowly, so the number of tampon applicators on a beach can represent years of accumulation and may become a rallying point for local activists. Figure 4-1 features a call to arms from local environmental activists to save the New Jersey beaches.*

Plastic tampon applicators are most commonly associated with scum, although they reach the beaches through a number of pathways. Aerodynamically shaped, with a diameter of roughly 0.5 in., most applicators slip through the screens at the

* TACKI is a project initiated by Jay Critchley of Provincetown, MA. Mr. Critchley is a performance artist who appeared around the country in a Statue of Liberty costume constructed solely of plastic tampon applicators found on the beaches of Cape Cod.

Table 4-1 Reported Scum Production and Properties

Treatment plant	Volume (ft^3/Mgal)*	Quantity (lb/Mgal)	Dry weight (Mg/l)	% Solids	% Volatile	BTU/lb Solids
N.W. Bergen Co., NJ	3.3	19.	2.3	9	—	—
3 New York City plants	0.04–0.7	1–15	0.1–2.0	40–50	—	—
Jamaica, New York City	3	10	1.2	—	—	—
Wards Island, NY	—	40	4.8	—	—	—
Passaic Valley, NJ	2.5	50	6	—	—	—
Chicago, IL						
Calumet	0.4	15	1.8	63	—	—
Southwest	1.2	44	5.3	61	—	—
West side	1.1	40	4.8	62	—	—
North side	0.3	4.5	0.5	53	—	—
Detroit, MI	—	25	3	—	—	—
Minneapolis-St. Paul, MN	—	—	—	—	98	5600
Milwaukee, WI	—	26	3.1	—	—	—
San Mateo, CA	—	95	11	—	—	—
Los Angeles County Sanitation District, CA	—	87	10	—	—	—
Sacramento, CA	—	120	14.4	—	—	—
East Bay MUD, CA	3.2	82	9.8	52	96	6000
Seattle, WA	6.7	24	2.9	6	—	
BP, St. Louis, MO	—	88	10.5	—	—	—
Albany, GA	—	140	17	0.6	—	—
Averages	1.8	49	5.9	40.1	97	5800
Range	0.04–6.7	1–140	0.1–17	0.6–63	96–98	5600–6000

Sources: Water Environment Federation[4] and Massachusetts Water Resources Authority.[3]

* Mgal = million gallon

Table 4-2 Summary of EP Toxicity Analyses Performed on Scum

Constituent	Detection limit[a]	Nut Island scum 1986	Detection limit[b]	Nut Island scum 1991	RCRA limit[c]
Arsenic	0.01	BDL	0.005	0.008	5
Barium	2.0	BDL	0.01	BDL	100
Cadmium	0.02	BDL	0.01	BDL	1
Chromium	0.2	BDL	0.050	BDL	5
Lead	0.2	BDL	0.001	0.007	5
Mercury	0.002	BDL	0.001	BDL	0.2
Selenium	0.01	BDL	0.005	BDL	1
Silver	0.02	BDL	0.03	BDL	5
Endrin	0.0001	BDL	0.0002	BDL	0.02
Methoxychlor	0.0001	BDL	0.0002	BDL	10
Toxaphene	0.0001	BDL	0.001	BDL	0.5
2,4-D	0.001	BDL	0.001	BDL	10
2,4,5-TP (Silvex)	0.001	BDL	0.002	BDL	1

Source: Massachusetts Water Resources Authority (MWRA).[3]

Note: All concentrations are in milligrams per liter. BDL = below detectable limits.

[a] Results of study by Haven and Emerson/Parsons Brinckerhoff, January 1986, at the Nut Island Treatment Plant, Boston, MA.

[b] Results of analyses performed by Matrix Analytical, Inc. on a sample taken 4/29/91 by MWRA.

[c] Limitations for EP toxicity characteristics of hazardous waste, 40 CFR 268.24 and 310 CMR 30.125.

Figure 4-1 Floating plastics can become a rallying point for local activists. (Courtesy of J. Critchley.)

headworks. The applicators are made of polyethylene, so they float in water, but in sewage their behavior is less predictable. Some of the applicators float and are collected with the scum, while others are carried along with the mainstream of the influent flow and discharged directly into the waterways with the effluent. If solids are trapped in the applicator cavity, they sink in the sedimentation tanks and are processed with the sludge. An additional source of plastic tampon applicators in the environment is from combined sewer overflows, when applicators will spill directly into the waterways with the sewage during each overflow event.[2]

4.2.3.1 Quantifying the Problem

The quantity of plastic tampon applicators that reach a POTW is very difficult to calculate. The following example illustrates why:

No. applicators used per day
= (service population × (% women) × (% potential users) × (% users) × (% of time used) × (no. used per person per day)
= 2,200,000 × 0.50 × 0.50 × 0.67 × 0.25 × 4
= 368,500 applicators per day

No. plastic applicators at the POTW
$= (368{,}500) \times (\%\ \text{plastic}) \times (\%\ \text{flushed})$
$= (368{,}500) \times (.60)\ (.10 \times .80)$
= 22,100–176,880 plastic applicators flushed per day

One half of the service population is female; one half of those women menstruate; two thirds of those women use tampons, which are used 1 out of every 4 weeks at a rate of four per day. Of the applicators used, 60% are plastic, and anywhere from 10 to 80% of the plastic applicators used are flushed down the toilet. The critical figure in this equation is the percentage of plastic applicators that are flushed. Tampons with plastic applicators clearly request on the box and on each individual wrapper that the applicator not be flushed. Nonetheless, every treatment plant operator knows that women flush applicators, although disposal rates may vary in different parts of the country. Screening of the influent flow in the Boston area indicated that roughly 50,000 plastic tampon applicators were reaching the treatment plant daily, which translates to a flush rate of about 40%. The Playtex Company, one of three companies in the U.S. that manufacture this product, claims that informal surveys show that only 1% of their customers flush the applicators.[3]

Scum and sludge are typically coprocessed and released to the digester; many of the plastic tampon applicators that entered the digester with the scum and sludge will float to the top of the tank and form a mat. Over time, this applicator mat can become as much as several yards thick and may compromise digester operations by reducing the active digester volume. In the past, when the digested scum and sludge were commonly discharged directly into local waterways or the ocean, the plastic tampon applicators that were not retained in the digester eventually washed up on area beaches or riversides. Now that the ocean dumping of sludge has been phased out and the beneficial reuse of sludge is more common, the issue of plastics in sludge is addressed by processing the sludge in such a way that plastics are not present in a recognizable form in the sludge product.

4.2.3.2 Plastics Reduction

The discharge of plastic tampon applicators from a POTW can be reduced in several ways. Source reduction to reduce the number of applicators flushed down the toilet is usually not possible, since tampon applicator disposal is a somewhat squeamish topic for public education. This leaves the option of removing the plastics from the waste streams by screening or reducing the size of the plastics in the scum and the sludge via grinding. Alternatively, scum can be disposed of separately from the sludge by landfill or incineration, which will remove most of the plastics from the sludge stream.

4.3 SCUM PRODUCTION

There are two types of scum produced at a wastewater treatment facility: raw scum and digester scum. The raw scum collected is only a fraction of the total grease entering the treatment plant. The typical grease content of raw domestic sewage is about 100 mg/l, with a range of about 25 to 150 mg/l.[1,4] When wastewater enters the primary clarifiers, some grease floats to the top of the tank, where it is collected as raw scum. Some of the grease adheres to the organic matter that sinks to the bottom of the tank, where it is collected as primary sludge. Finally, a substantial fraction of the grease exits the POTW with the effluent flow. Since the grease may be diverted to the sludge, scum, or effluent, scum production does not directly correlate with influent grease loadings or with the removal of suspended solids at the treatment plant. Likewise, although raw scum is largely composed of grease, it is routinely an insignificant carrier of the total grease loading at a wastewater treatment plant. The partitioning of influent grease and oils in a 300-MGD treatment plant is shown in Figure 4-2. Typically less than 10% of the grease entering a treatment plant is collected as raw scum, a figure which varies with the specific skimming procedures and reporting techniques at each wastewater treatment facility.

4.3.1 Scum Collection

During primary treatment, scum is collected from the surface of the sedimentation tanks or skimming tanks. Scum is usually collected at the effluent end of the primary sedimentation tank. Some plants, however, have located the scum collection site on the influent end of the sedimentation tank to decrease the travel distance of scum to the collection point, ensuring rapid removal of all flotage. Influent scum collection on a primary sedimentation tank is shown in Figure 4-3. Preaeration, which enhances scum separation in the primary sedimentation system, used to be a routine operation. Now that secondary treatment is the rule rather than the exception, preaeration is no longer as common.

Automated scum removal devices may be attached to the sludge collection mechanism, or they may be operated separately. The inlet to the sedimentation tank should be designed to allow scum to enter freely, without building up on the inlet channels or behind baffles. The scum removal mechanism should extend the full width of the tank to prevent floating material from reaching the effluent weir. If the influent is particularly greasy, a separate skimmer mechanism or water spray system may be necessary. Maintaining a constant level in the sedimentation tanks improves the performance of scum collection mechanisms.[4]

For small installations the most common scum drawoff facility consists of a horizontal slotted pipe that can be rotated with a lever or screw: a tilting trough scum collector, shown in Figure 4-4. The open slot is kept above the water level except during scum collection, when a tripping device on the sludge collector rotates the trough to permit the scum accumulation to flow into the trough and then to a wet well. This process results in a relatively large volume of scum liquor.

The sloping beach, shown in Figure 4-5, is a stationary device with a collector trough. Floating material is directed to the sloping beach with jet air sprays, water jet sprays, the sludge collection mechanism, or a separate blade-type scraper. A transverse rotating helical wiper attached to a shaft can draw the scum from the water surface over a short inclined apron for discharge into a cross-collecting scum trough. The scum is then flushed through the trough to a wet well from which it is pumped to another treatment process.[4]

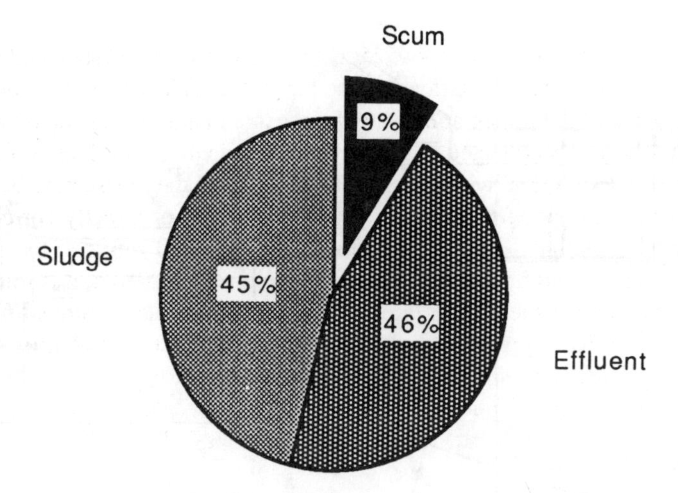

Figure 4-2 Grease partitioning during primary treatment.

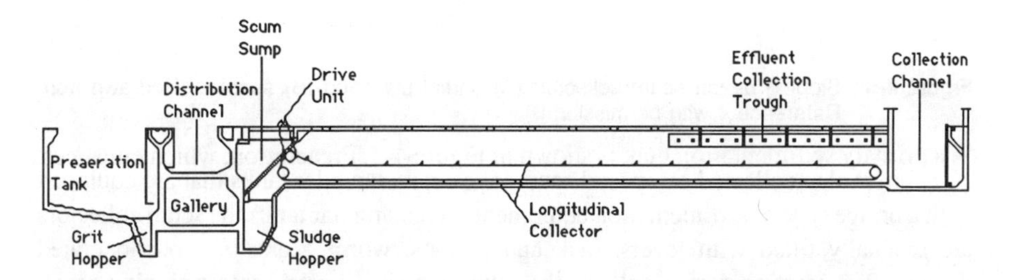

Figure 4-3 Primary sedimentation tank with influent scum collection. (Redrawn from Reference 4 with permission.)

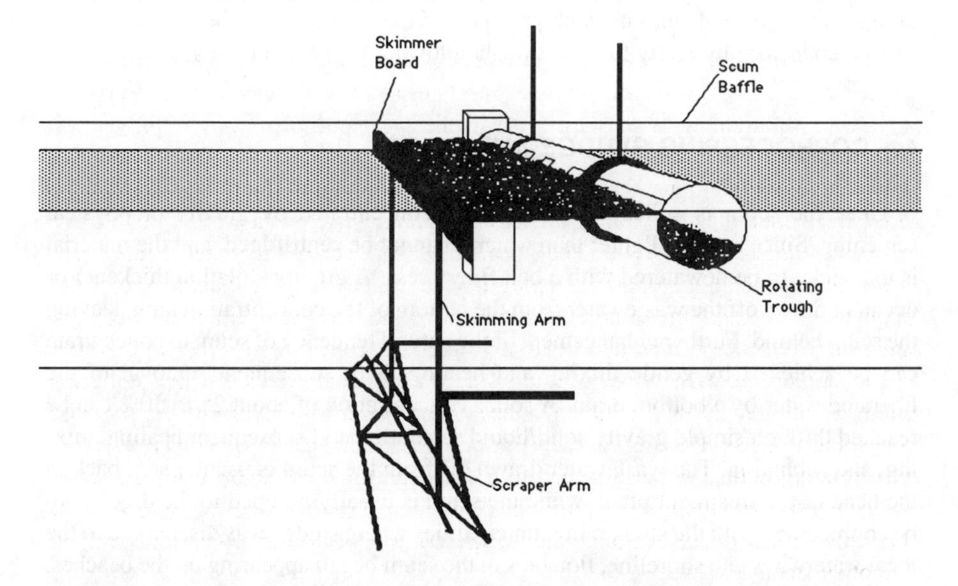

Figure 4-4 Tilting trough scum collector. (Redrawn from Reference 4 with permission.)

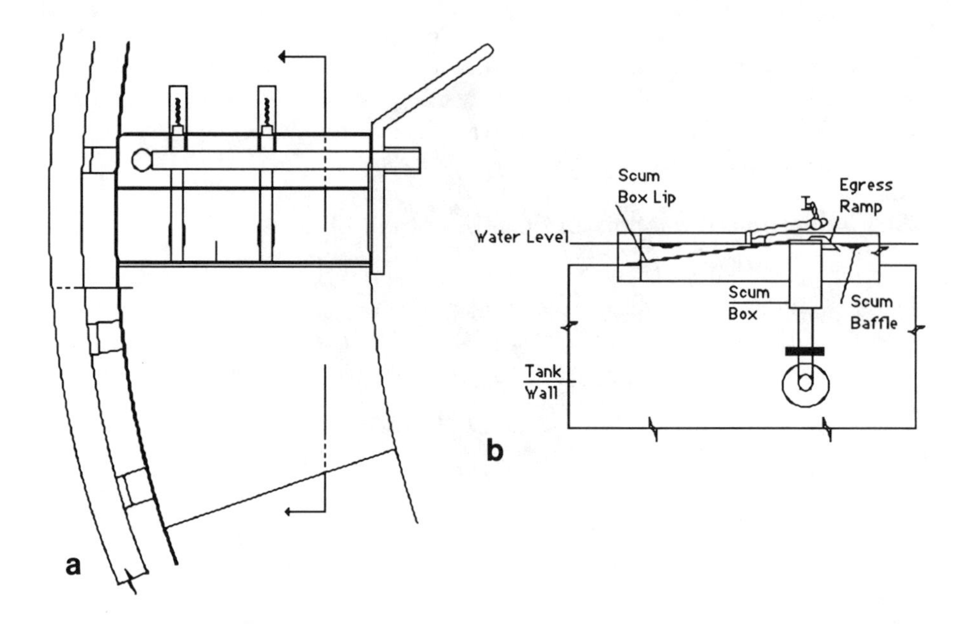

Figure 4-5 Sloping beach scum collector: (A) partial plan and (B) section. (Redrawn from Reference 4 with permission.)

Scum is also collected by special scum rakes in rectangular tanks that are equipped with a bridge type of sedimentation equipment. In rectangular tanks the scum collectors are manually tilted with levers, rack and pinions, worm gears, or motor-operated devices. The rotating device collects the scum, which is pushed into a sloping beach or a tilting trough skimmer. A spring-loaded section of the rotating arm rides up the beach, wipes scum into the trough, and drops into the water on the far side. The tilting trough usually extends into the tank just short of the rotating arm; the passing arm or surface collector physically tilts the trough, allowing it to collect the scum.[6]

4.4 COPROCESSING SLUDGE AND SCUM

Once the scum is collected, it is usually concentrated by gravity or physical screening. Since scum is lighter than water it cannot be centrifuged, and the material is too sticky to be dewatered with a belt filter press. A gravity-flotation thickener or decanter draws off the wastewater from the bottom of the concentration tank, leaving the scum behind. Further enhancement of the natural tendency of scum to concentrate can be achieved by gentle mixing and heating, with subsequent removal of the liberated water by a bottom drain. A solids concentration of about 25 to 50% can be reached through simple gravity solid/liquid separation and subsequent heating, mixing, and subnating. The wastewater drawn off from the scum is usually sent back to the head of the treatment plant, while the scum is usually pumped to the digester to be coprocessed with the sludge. In municipalities where sludge was discharged to the local waterways and shoreline, flotables in the scum began appearing on the beaches.

In sludge that is beneficially reused, plastics are equally unappealing. Grinding the scum and sludge reduces the particle size to the point where flotables are unrecognizable. Screening the residuals removes the flotables from the reusable material, and the plastics are ultimately landfilled.

4.4.1 Anaerobic Digestion and Gas Formation

In the digester, the biodegradable grease in scum is broken down by anaerobic digestion. During the anaerobic process, organic matter is biologically converted into methane (CH_4) and carbon dioxide (CO_2). This is governed by two groups of bacteria: facultative anaerobic producers that convert carbohydrates, proteins, and fats into organic acids and alcohols, and aerobic fermenters that convert the acids and alcohols into methane and carbon dioxide.[6]

The gas produced by the breakdown of different substances in an anaerobic digester was quantified a half century ago. Table 4-3 shows the value of scum digestion for methane production.

Methane gas production and chemical oxygen demand (COD) reduction (as opposed to volatile solids reduction) are dependent on the anaerobic digestion of the biodegradable fraction of the grease entering the digester. Grease usually accounts for less than a quarter of the volatile material in the influent flow. About 80% of the grease is saponifiable, which means that it is solubilized under mixing with hot alkali. The destruction of this saponifiable fraction contributes over 50% of the total methane gas production and COD reduction during sludge processing.[5]

The contact time in the digester appears to be the key consideration in grease destruction. Bench-scale tests indicate that grease requires a minimum contact time of 8 to 10 days before it begins to break down. If COD reduction or gas production is to be maximized, scum degradation may be the rate-limiting step in anaerobic digestion.

Figure 4-6 is the result of bench-scale digestion, where the digester mixing is true and there is no accumulation of grit and scum to reduce the active volume of the digester.[3] Real digesters with less than ideal mixing have a layer of grit, a layer of scum, and areas of incomplete mixing, so the sludge needs an appreciably longer detention time for complete digestion of the grease.

Most digesters empty through the bottom, and scum floats. The plastic tampon applicators entrained in the scum and sludge that is pumped to the digester quickly separate and float to the top; they rarely exit by a port at the bottom of the digester. The plastics provide a matrix for the layer of digester scum, which may build up to 6, 8, or even 10 ft thick. At this point the scum layer begins to impede digester performance, and remedial steps must be taken. Most plants periodically use gas recirculation systems to accelerate scum digestion and reduce the depth of the scum layer. Digesters are routinely emptied a few times a decade to clear out the scum and grit accumulations.

4.4.2 Digester Mixing

The scum layer in the digester will be much less significant if the digester is well mixed. A well-mixed digester

Table 4-3 Sources of Methane in Sludge

Sludge feed	Gas produced per pound of volatile material destroyed (ft³)	% Methane
Fats (lipids)	18–23	62–72
Crude fiber	13	45–50
Protein	12	73

Source: Buswell and Neave.[10]

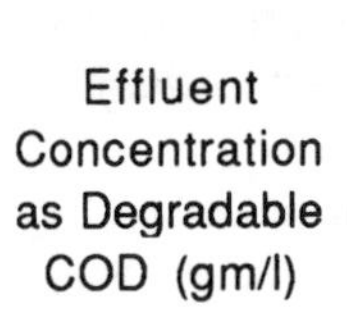

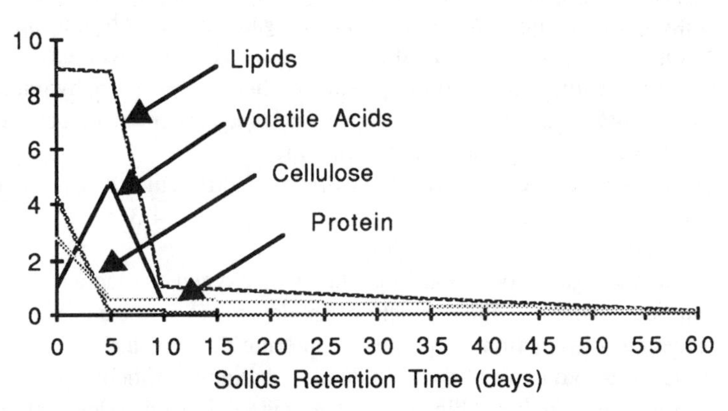

Figure 4-6 Bench-scale digestion of primary sludge at 95°F (35°C).

1. Maintains contact between active organisms and the digester contents
2. Minimizes concentrations of biological inhibitors
3. Maintains the active digester volume by impeding the formation of a scum layer and the deposition of suspended solids on the bottom of the tank

All of these factors reduce scum production.

Conventional digesters can be mixed by mechanical methods or by gas recirculation. Mechanical mixing requires an energy input of about 0.25 hp/1000 ft³ of digester volume. "Weedless" propellers that are specially designed to avoid fouling by the raggy, inert nonfluid material or flat-bladed turbines can be either top or side mounted. Gas recirculation is employed in several variations. A large sludge gas bubble can be injected into the digester to create pumping action, gas can be injected by a series of lances hung from the digester cover, a draft tube inside the digester can be used for the unconfined release of gas, or a floor-mounted ring of diffusers also can be used for the unconfined release of gas.[3]

The injection of a sludge bubble into the digester has a low power requirement and does not produce much mixing. Lance and draft mixing produce considerable bottom velocities; this reduces the accumulation of settleable material. Free gas lift systems have a bottom bubble velocity of zero which accelerates to a maximum at the digester surface.

Aerobic and anaerobic digesters are widely used to partially digest scum that is coprocessed with wastewater sludges, but scum digestion is problematic. In aerobic digesters, grease balls may form; the inclusion of scum may result in petroleum contamination of the sludge, reducing options for reuse. The appearance of the resulting sludge may be degraded, and scum will build up in the digester if the remaining residue is not completely removed. Anaerobic digesters have the same problems as aerobic digesters, and a few others as well: if the digester is not strongly mixed, a scum layer is certain to form. In general, conventional digesters have "dead spots" in which solids accumulate, causing poor digestion and periodic shutdowns for cleaning. The relatively large surface area allows a scum blanket to form. This blanket is difficult to remove without emptying the digester, and it is difficult to dispose of after cleaning. In addition, the scum must be well decanted before introduction to the digester to avoid adding significant quantities of water to the digester when the scum is applied. Unfortunately, the less water entrained in the scum, the more likely the lines are to clog with congealed scum. Egg-shaped digesters reduce scum production by eliminating dead spots in the digester and by providing a smaller surface area for the scum blanket to establish itself, as shown in Figure 4-7. With 3 installations in the U.S. and over 100 in Germany, these ovoid digesters are clearly an improvement in digester design. They provide much better mixing of digester contents, reducing scum production and making the grit layer of conventional digesters obsolete.

4.4.3 Grinding

When sludge and scum are ground before being discharged to the digester, the plastic applicators are no longer recognizable as "beach whistles". If the principal concern is aesthetics, then making the applicators unrecognizable is an effective strategy. Grinding facilitates mixing and intimate contact during the digestion process and should enhance grease destruction and gas production. The streamlined conformation of the applicators makes grinding difficult; two-stage sludge grinding may be necessary.

4.4.4 Screening

Screening the sludge and scum can effectively remove most of the plastic applicators from the waste stream. Screening separates particles as a function of the size of the opening of the screen. Smaller openings yield a more demanding unit operation and provide a greater opportunity for odor evolution due to the increased use of cleaning steam and hot water to remove the scum that builds up on the screen and causes blinding. The high-temperature washings result in a flux of volatile or strippable organic and inorganic compounds that smell.

The 410-MGD Hyperion Plant in Los Angeles screened its sludge for over a decade before finally stopping disposal of sludge in the Pacific Ocean. In total, the plant had 19 static Bauer screens, each with a 0.06-in. screen opening and rated at 100 to 200 gal/min/unit; 11 were used for routine sludge screening, while the other 8 were used for digester cleanouts. The operations included sludge screening units, screening reject conveyors, and belt filter dewatering (without polymer) of the reject to about 20 to 25% solids. The dewatered reject was stored in hoppers, ground, and reintroduced into the anaerobic digesters. The dry weight of the reject mass was reported as about 5% of the feed sludge; it probably contained little scum since it was successfully dewatered with a belt press. Although the scum and sludge were

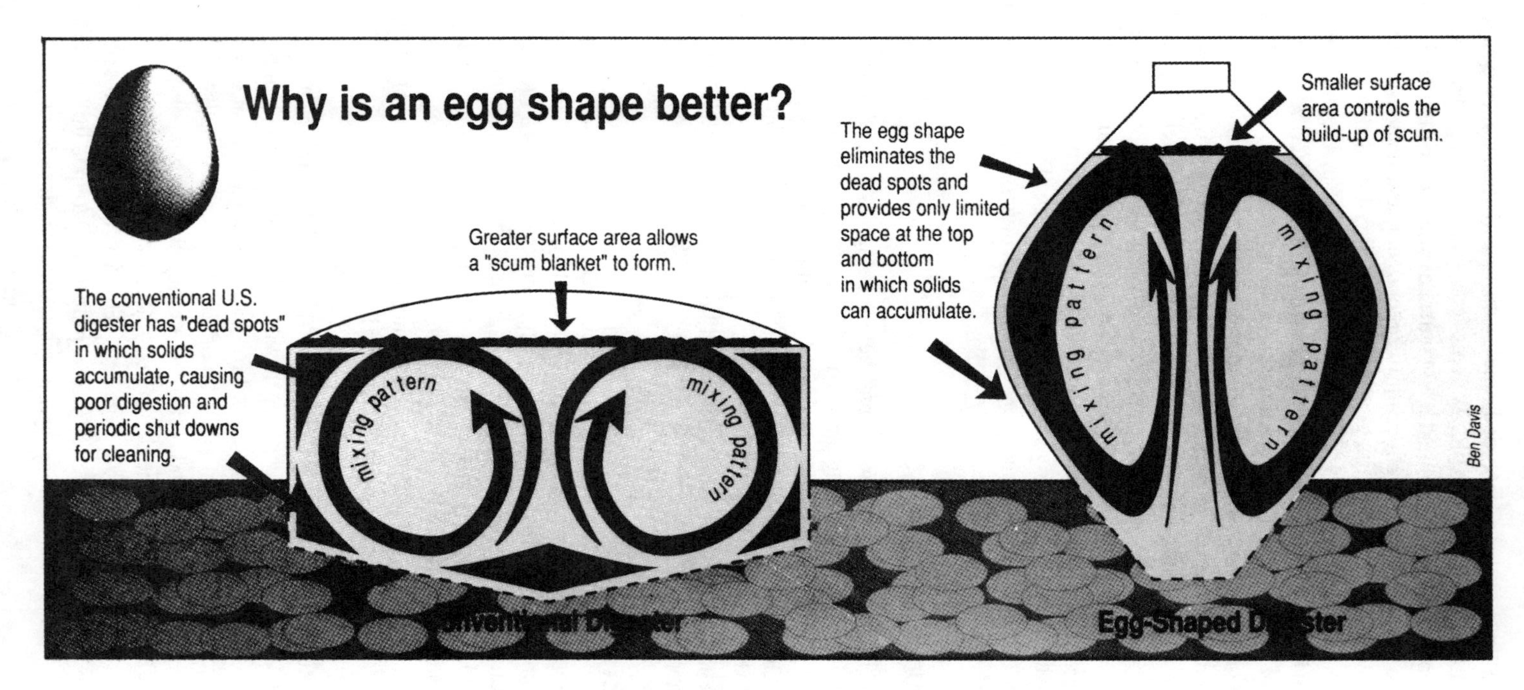

Figure 4-7 Egg-shaped digesters reduce scum production. (Courtesy of MWRA)

screened to prevent the release of flotables, the installation appeared to remove a variety of unrelated constituents.[3]

Hyperion has since discontinued sludge and scum screening and reverted to discharging its scum to the digesters for coprocessing and disposal with sludges.

4.5 SEPARATE PROCESSING OF SLUDGE AND SCUM

If the sludge shows high levels of petroleum hydrocarbons or exotic organics, these substances may be partitioned into the scum before it is discharged to the digester. If so, one option to improve sludge quality would be to process scum and sludge separately. Separate processing of scum and sludge reduces the amount of methane generated during digestion. Also, scum disposal options are limited: it can be landfilled, incinerated, or beneficially reused.

The landfilling of scum traditionally has been an alternative with low capital costs, although operating costs may be relatively high. The scum must be well decanted to minimize volume and fluidity, for the heavy equipment can slip on the grease. If separate processing and landfill disposal is the preferred option, the material may have to be lime stabilized before landfilling. The odors and infectious nature of scum, coupled with the difficulty of handling it, may make unlimed scum unwelcome at municipal landfills.

Incinerators are expensive both to construct and to operate. Siting the facility may become highly political, and operations can be plagued with air quality control problems. Scum incinerators should be run continuously to maximize furnace performance and minimize air quality control problems. The high capital and operating costs make incineration an unacceptable solution for small plants. The U.S. EPA has suggested that scum be beneficially reused as animal feed or in low-grade soap manufacture. There may be isolated instances where this concept has proved to be viable, but since most toxic organics concentrate in the grease and there are better, low-cost alternatives, it is an option of doubtful attractiveness. The beneficial reuse of scum is, therefore, for practical reasons limited to chemical stabilization to reuse the scum as structural fill.

4.5.1 Incineration

In 1939, the Minneapolis-St Paul Sewage Treatment Plant disposed of its scum in two lagoons that were alternately filled and set afire. It is amazing to realize that this simple, economical solution to scum disposal was practiced for nearly 25 years.

The incineration of scum is no longer so casual. The heat value of scum is so high that air quality control problems and flare-ups are common. Sludge particulate emissions are much easier to control than scum emissions. The difficulty of implementing new air pollution control and source performance standards effectively limits scum incineration to fluidized bed incineration. The most successful of the demonstrated incineration strategies are those which only handle scum and are sized for continuous operation, reducing the air quality problems associated with start-up, shutdown, and varying feedstocks.

Nichol's WATERGRATE® system for scum incineration was very favorably referenced by the EPA in 1979 in their Technology Transfer Manual and was beginning to gain widespread acceptance when a severe accident in 1981, involving a start-up explosion with higher than anticipated BTU feedstock, resulted in the complete abandonment of this technology.[3]

In fluidized bed incineration, a fluidized bed of sand is used as a heat reservoir to promote uniform combustion. Most fluid-bed combustion systems are designed and operate at combustion temperatures between 1500 and 1650°F (815 to 900°C), with preheated air temperatures between 900 and 1200°F (482 to 650°C). The convection fluidized bed combustor is suitable for incinerating scum because of its inherent good mixing, long retention times, and ability to be operated at a low excess air ratio of 40%. Afterburners are routinely used to maintain air quality standards.[8]

The high capital costs of incineration technology and the potential for public demand to close down a furnace make it unlikely that an incineration technology other than fluidized bed incineration will gain in popularity.

4.5.2 Chemical Fixation

Liquid scum can be chemically stabilized and converted into a pasteurized, odorless, soil-like material by mixing it with a material with a high pH and a very fine particle size, such as cement kiln dust, lime kiln dust, or ground lime. Cement kiln dust is the by-product of cement manufacture and is comprised primarily of calcium and magnesium oxides and silica.

At the Deer Island Treatment Plant in Boston, scum was chemically fixed with cement kiln dust and used as structural fill. Scum was mixed with cement kiln dust to a slurry in a cement mixer and transported to a curing basin. The ratio of cement kiln dust to scum depends on the water content of the scum and the dust, the temperature of the curing pits, and the length of time allowed for curing. Typically, 20 to 75% cement kiln dust by weight is added.

After discharge into the curing basin, the cement kiln dust and the scum begin to react. The high pH of the cement kiln dust effectively kills all pathogens, and virtually all odor is eliminated. Some free water is released and periodically returned to the head of the treatment plant. The remaining scum and cement kiln dust combine to form a structurally sound, soil-like material. When the mixture reaches this stage, it is excavated by backhoe, trucked, and stockpiled using conventional soil handling equipment. Depending on the season and the amount of cement kiln dust used in the mixture, scum is effectively stabilized and solidified within 1 to 4 weeks of initial mixing. Quicker "set-up" time is achievable, but not economically warranted.

100 million kg (100,000 t) of chemically fixed scum was applied in the District of Columbia from 1981 through 1984 for landfill cover, and 170 m^3 (400,000 yd^3) was used in Wilmington, DE for berms and embankments.[9] Oily wastes were also successfully solidified by chemical fixation. Principal complaints included improper curing, too low concentrations of solids, and the release of ammonia upon mixing with the sludge because of the high pH of the mix. With attention to product quality control, chemical fixation is a viable scum processing alternative.

REFERENCES

1. "Sludge Treatment and Disposal Process Design Manual," U.S. EPA Technology Transfer, Municipal Environmental Research Laboratory, EPA 625/1-79-011 (September 1979).
2. Outwater, A.B. "Degradable Plastics and Wastewater," Meeting of the Massachusetts Marine Bay Consortium in Boston, MA (November 1988).
3. "Scum Disposal Feasibility Study: Technical Support Document," Prepared for the Massachusetts Water Resources Authority Sewerage Department by Havens and Emerson/Parsons Brinkerhoff (March 1986).
4. "Design of Municipal Wastewater Treatment Plants, Vol. 1," Manual of Practice No. 8, Water Environment Federation, Alexandria, VA (1992).
5. Harris, R.H. "Skimmings Removal and Disposal Studies," Research and Development Section, Los Angeles County Sanitation District, Los Angeles, CA (September, 1964).
6. Metcalf & Eddy, Inc. *Wastewater Engineering: Treatment, Disposal, and Reuse* (New York: McGraw-Hill, 1972).
7. Davis, B. "Why is an egg shape better?" *Water Resources,* 3(4):4 (May/June 1990).
8. Coker, C.S., R.L. Walden, T.G. Shea and M.J. Brinker. "Dewatering Municipal Wastewater Sludge for Incineration," *Water Environ. & Technol.* (March 1991).
9. Mulbarger, M.C., R.P. Trubiano, T.L. Papes, and G.L. Gallinaro. "Past Practices and New Approaches in Scum Management," *Water Environ. & Technol.* (November 1990).
10. Buswell and Neave. "Laboratory Studies of Sludge Digestion," Illinois State Water Bulletin 30 (1939).

5 DESIGNING A SLUDGE RECYCLING PROGRAM

5.1 INTRODUCTION

To beneficially reuse sludge, a market must be created for a sludge product. In most cases, a public aversion to municipal sludge requires that some level of sludge processing is carried out to produce a marketable product. In a successful sludge recycling program, a sludge processing option that suits the individual economic and geographic limitations of the POTW is paired with a market that meets the environmental and reliability requirements of the facility.

A wide variety of sludge products can be produced by a municipal wastewater treatment plant. Sludge can be land applied as a liquid, spread as sludge cake, or pelletized; it can be composted, incinerated, or stabilized with additives to be reused as structural fill. The range of markets for sludge products is even broader. Processed sludge can be used by the boxcar to reclaim strip-mined lands; it can be sold in garden centers in 25-lb bags. It can be spread in forests or in flower gardens, sold to topsoil dealers, or used as final root zone cover for landfill closures. There are so many combinations of products and markets that narrowing the field of possibilities down to a coherent sludge reuse program is a complex proposition. Before deciding which sludge product and market is the most appropriate solution for your community, it is important to take a critical look at some of the underlying factors.

5.2 FACTORS AFFECTING SLUDGE REUSE

Cost is an important factor influencing the method of sludge processing chosen by a community, but it is not the only important issue. There are a variety of emphases which are used to select the most appropriate method of sludge reuse for a specific wastewater treatment plant. These factors must be explicitly defined and ranked before a community can reach an optimal solution for a sludge recycling program. Table 5-1 provides a list of issues which should be rated as important, moderately important, and relatively unimportant. This information provides a sketch which can be used to prune the field of processing options and markets.

Table 5-1 Factors Affecting Sludge Reuse

Economic	
Cost per ton	Operational costs including sludge processing, hauling, application, and site monitoring
Present worth	Cost in today's dollars over the life of the project, including revenues and capital costs
Revenue/benefit	Monetary or other economic benefits to the POTW or others, short or long term
Avoided costs	Avoided costs of sludge disposal, such as landfill tipping fees
Cost of delay	Time constraints may be structured as fines or grant limitations
Geographic	
Project location	Within the service area, or in a political jurisdiction friendly to the sludge project
Land use	Compatibility with local land use, regional plans, and zoning regulations
Site control	Site owned and permitted; access, degree of buffer, fencing, and storage provided
Size of site	Land pressures in urban areas may limit the size of available processing sites
Land values	Cost and availability of processing sites
Sludge transport	Relative rating of haul distance, route, or mode of transportation; traffic impacts
Nuisance factors	Potential for odor concerns, traffic issues related to hauling, or aesthetic impacts
Environmental	
Water quality	Site-specific risk of ground- or surface-water pollution; application rates and land use
Terrestrial habitat	Potential for disturbance, displacement, or contamination through processing or end use
Human health	Risk of exposure to humans directly or via the food chain; risk of litigation
Reliability	
Experience	Prior experience/credibility of project contractor; tried and true processing option
Flexibility	Multiple options for processing and marketing
Simplicity	Ease of process operation
Acceptability	Effort required to sustain public acceptance and awareness of sludge recycling
Public service	Project meets identified public needs
Other	
Expandability	Ability of process to accommodate increased waste streams
Specificity	Capacity to accept other wastes and maintain good end product
Cost sensitivity	Reliance of processing costs on fuel or chemical products

Source: Seattle Metro.[3]

5.2.1 Economic Factors

Economic factors include the operational costs of sludge processing, the lifetime costs of the processing facility, the revenue potential of the sludge product, the avoided costs of alternate disposal, and the cost of delay. If the total costs are the critical factor, longer-term benefits and revenues of a project may offset higher initial costs. Long-term benefits may include such assets as direct revenue, harvestable timber, or monetary credits that reimburse part of the initial investment. The cost of capital, either short term or long term, may determine the sludge processing option chosen, restricting options to processes that require minimal capital investments or weighting those options with high capital costs and low operating expenses. If there are no available landfills or if tipping fees are exorbitant, the avoided costs of disposal accrued by a sludge reuse program may be substantial. Conversely, if there is an underutilized, inexpensive incinerator nearby, then sludge reuse may be a relatively costly undertaking. If public opposition to the project could cause delays that would add significantly to the overall cost of the project through fines levied against the current disposal method or by increasing the cost of capital, then expenditures on public education may be an important part of the sludge reuse program.

5.2.2 Geographic Factors

Political realities may limit available processing sites to parcels within the collection area of the POTW or within communities sympathetic toward the sludge producers. A sludge processing site must be compatible with local land uses, regional development plans, and zoning regulations. The municipality may own a permitted site with a given buffer zone and storage area, presenting the POTW with a *fait accompli* that limits available processing options. Urban communities may be subject to siting pressures which preclude the possibility of choosing a processing option that requires a larger land area, such as windrow composting. Land pressures also may preclude any available site with neighbors near enough that odor control becomes of paramount concern. Land values may limit the availability and raise the cost of processing parcels.

If sludge is to be applied directly to farmland in a rural location, the proximity of a project to the wastewater treatment plant or specific location within a service area may determine whether the sludge is applied as a liquid or spread as sludge cake. When a sludge processing facility is sited separately from the POTW, the sludge may be transported between facilities by barge, truck, or pipeline. Nuisance factors usually concern the proximity of a processing site to the nearest neighbors or concerned citizens group. These include odor issues, truck traffic, and aesthetic impacts of the processing facility.

5.2.3 Environmental Factors

The end use for a sludge product is usually the most important determinant of the environment impacts of a program. Water quality impacts include the risk of groundwater contamination and impacts to local waterways from sludge-contaminated surface runoff. Most water quality issues are site specific, relating to local soil conditions, slopes, and weather patterns. Application rates that are designed to protect human health and the environment are presented in loadings per year. Water quality issues arise from the fact that sludge is applied in step intervals: in real time a site does not see an annual loading assimilated over the course of a year; instead, sludge is applied as a slug loading. Newly applied sludge products are liable to be washed into the waterways by a sudden rainstorm or leech into the groundwater shortly after application. These problems are addressed by monitoring application rates, sludge solids content, and land use practices. Sludge that is injected directly into the soil will have less of an impact on water quality than sludge that is sprayed on a site as a liquid.

Terrestrial habitat impacts can be critical when sludge is used for silviculture or, occasionally, agriculture. When large quantities of liquid sludge are applied at one time, the potential for environmental disturbance, displacement, or contamination may be significant.

5.2.4 Reliability

For a sludge reuse program, reliability has a number of different guises. Reliability of the sludge processing technology may be critical; companies with a proven track record and a standard technology would be most attractive. Flexibility in sludge processing options, where a single POTW may use a few different processes to make sludge products for a variety of markets, can provide the certainty of product dispersal that a wastewater district requires. Most of the larger cities in this country use more than one sludge processing technology to provide program flexibility. The ease of operating a sludge processing facility may be a factor, as well as the ease of distributing the sludge product. In other areas, public service initiatives like green space enhancement or land reclamation may be the deciding factor. When sludge products are distributed in end markets that are chosen for their visibility, the program maintains public awareness and support for sludge recycling. If the sludge products are destined for highly visible markets, processing options may be limited to pelletization or composting to produce an end product that doesn't smell like sludge.

5.2.5 Other Factors

The possibility of expanding a sludge processing operation to accommodate increases in sludge production may be an important factor. Some municipalities choose sludge processing options that can accept multiple waste streams. Yard waste programs can be paired with sludge composting, lowering costs for both. A sludge pelletizing facility can also accept restaurant grease and veterinary wastes (dead cats and dogs) with no degradation of the finished sludge pellets. Finally, processing options that require a great deal of energy (such as pelletizing or horizontal in-vessel composting) create a level of uncertainty in future energy costs that may be unattractive to some reuse programs.

5.3 SITING SLUDGE PROCESSING FACILITIES

If it were easy to find sites to land apply sludge it wouldn't be necessary to build sludge processing facilities. The difficulty of finding acceptable sites for liquid sludge application has given rise to a variety of sludge processing technologies that make a sludge product appealing enough to be marketed. This significantly changes the problem of siting: instead of trying to permit and monitor sites for land application, a permanent site for a sludge processing facility must be secured. The problems of siting a facility may be a significant factor in choosing a sludge processing technology.

Beneficially reusing sludge is a task with a far higher public profile than simply disposing of the material. When a POTW sites a sludge processing facility, it may be the first time that the public is aware that their toilets connect to a wastewater treatment plant. People that didn't know sludge existed a week ago are suddenly deeply concerned about heavy metals, environmental impacts, and odors. In the past three decades the nuclear power industry has gone from "energy too cheap to meter" to a virtual standstill; this can be largely attributed to the poor response to public concerns by the U.S. Nuclear Regulatory Commission (NRC). This has created the perception that large public entities are monolithic, unresponsive, and irresponsible and that they must be stopped. In some areas of the country a sludge facility can still be sited with little fanfare; in others there may be a strong community perception that a sludge processing facility is a nuisance that will decimate land values, pollute the environment, and destroy the quality of life. The reality is that if a community opposes a siting decision, the cost of delay can make the situation untenable for a POTW: opposition can and does kill sludge reuse projects.

It is now recognized that the recent advances in dispute resolution and mediated agreements provide the formula for successful siting. The basic premise of siting theory is that the public should be included in the decision-making process. The site is chosen by building a consensus between the community and the POTW, so the public is not presented with a final solution. Instead, they are collaborators in the process.

5.3.1 Site Selection Criteria

The first step in the siting process is to meet with the opinion leaders of the community to establish the criteria for evaluating a site. Opinion leaders of the community are not necessarily the people who are outspoken about an issue; they are the people with long-term standing in the community who are recognized and respected. Local nonprofit environmental groups may be perceived to be an unbiased, educated resource with an informed perspective untainted by monetary concerns, and their objections also should be carefully addressed. The public meetings serve to educate the community so its constituents are able to make informed decisions, as well as to provide a forum for public concerns. The meetings should be held in small groups. It may take many meetings before all the concerns are heard.

It is critical that the criteria are agreed upon *before* any site is selected. This step is often omitted. Project sponsors often have a site in mind and establish criteria that meet that location without involving citizens at an earlier stage in the process. The bald facts are that the community pays for the facility, and individuals who want a say in how the project is implemented must be educated through public meetings so they can be an asset to the process rather than a roadblock. Once criteria are formed by consensus, the site selection process can be evaluated openly. Common siting criteria are shown in Table 5-2.

5.3.2 The Environmental Impact Statement Process

As soon as the site and technology are chosen, the environmental evaluation begins. The opinion leaders are brought in again, and another round of meetings should be held to ensure that the environmental evaluation addresses areas of public concern. Environmental impact statements (EIS) are often done according to a regulatory protocol that may not respond to the specific concerns of a community.

Table 5-2 Common Siting Criteria for Sludge Processing Facilities

Proximity to the POTW	Community support for project
Land values	Compatibility with other planned uses
Site size	Visual impact
Development costs	Distance from neighbors
Existing utilities	Possible building reuse
Access to major roads	
Rail access	

Source: Adapted from Goldstein.[1]

For example, an EIS study of surface runoff may not address the community concerns of how to prevent hazardous wastes from entering the waste steam and what to do if this does occur.

Assumptions should be spelled out. If traffic is a concern, it is important to determine if it is rush hour traffic that the community objects to, or weekend traffic, evening traffic, road degradation, odor, noise, or something else. There has to be another round of education when the EIS process begins because public participants need to understand the methodologies used for testing. Once people appreciate how air quality impact is derived and what goes into the computerized models, the level of trust in the study results is much higher. The details behind a worst case scenario are also important. Once people understand the conservative numbers in these evaluations and appreciate how realistic it is that the worst case should occur, confidence in the facility builds quickly.

The EIS process is long, and it is important not to create an information vacuum during this period. The public participants, who are now educated in the arcana of sludge processing options and have helped select the site and processing technology, are eager for a continued dialogue. The painstaking process of developing an EIS may leave the POTW with nothing more to say than "We don't have the data yet." Don't say this. This period should be used to pass on information and experience from similar projects in different locations, for if information is not provided during the time that the EIS is being completed misinformation will fill the vacuum.[1]

5.3.3 Siting Pitfalls

If the project sponsors feel they must have a site in mind before going to the public, the process can fall apart in the beginning. The POTW is immediately in the position of having to defend its choice. Likewise, if a completed plan is presented to a large group, the meeting may become an opportunity to vent anger. This can create the sort of information program that will kill a sludge reuse project, for the media tend to report the most sensational statements made at public forums, and the public will get the impression that the project is poorly conceived and that most people are against it.[1]

5.3.4 Compensation

There is a stigma attached to sludge processing that overshadows specific concerns such as odors, truck traffic, or property devaluation. No community likes to be seen as a dumping ground, and one way to overcome opposition is to acknowledge this by offering benefits as compensation for the detriments of hosting a sludge processing facility. Compensation becomes an attractive option when there are few parties to the dispute, the opponents are well organized and geographically defined, impacts are clearly defined, and all parties are capable of offering a binding commitment. It is important to keep in mind that mutually acceptable outcomes do exist. Compensation may include a patchwork of goodies, as listed in Table 5-3.

Specifying the appropriate type and amount of compensation is not straightforward. The parties involved in the development of a sludge processing facility should negotiate the compensation package directly. The compensation agreement made by the Massachusetts Water Resources Authority and the City of Quincy, where an enormous sludge processing facility was sited, included annual cash disbursements of $2.25 million, job training programs, and compensation for fire protection services and traffic services. In addition to defining specific measures to mitigate the effects of sludge processing on the community, the agreement provides $100,000 a year for the host city to hire engineers and lawyers to watchdog the design, construction, and operation of the facility. Steps for dispute resolution are clearly defined, as are extraordinary circumstances.[2] This skillfully crafted compensation package has allowed construction of the largest sludge pelletizing facility in the U.S. to proceed without delays and with continued good relations with the host community, in spite of a highly politicized facility siting process. Compensation agreements can and do work.

5.4 SLUDGE PRODUCTS

Sludge can be processed to produce a range of products suitable for land application. The critical differences between these products lie in their water content, fertilizer value, marketability, and aesthetic impacts. Digested liquid sludge, at 1 to 3% solids (or thickened up to 10% solids), has the lowest fertilizer value per volume of sludge products, but can be reused without further processing in some situations. Dewatered sludge cake has far less volume than liquid sludge and is easier to transport. It is still, however, a relatively unappealing product. Sludge digestion, lime

Table 5-3 Compensation for Host Communities

Money	To local governments that could then reduce tax rates or increase services to residents in the form of direct payments
Conditional compensation	When some costs of development are possible but not certain, such as property value losses, a municipality might guarantee property values or offer other types of insurance as forms of compensation
In-kind compensation	A municipality might acquire land and develop it as a park to balance the loss of land that will be used for the sludge processing facility
Protection	Mitigating health and safety impacts by providing specific protections, such as a new fire truck or close monitoring of the facility
Impact mitigation	Reducing or eliminating negative impacts directly, such as by resurfacing roads which will be impacted by facility construction, improving odor control, or using water transportation for sludge and sludge products

Source: Adapted from O'Hare, Bacow, and Sanderson.[4]

stabilization, and air drying are considered to be processes to significantly reduce pathogens (PSRPs), and low levels of pathogens present make the class B/PSRP sludge products unsuitable for some applications. Further stabilization is accomplished by heat drying, composting, irradiation, or pasteurization, known as processes to further reduce pathogens (PFRPs). Pathogens in sludge products made by PRFPs are below detectable levels, creating a class A sludge that has wider marketing options. Lime-stabilized sludge can be applied to acidic fields or disturbed land or used as daily landfill cover. Composted sludge cake is a nonoffensive material that has little fertilizer value; instead it is considered to be a soil amendment. Sludge pellets are relatively costly to produce, but at 95% solids are relatively inexpensive to transport. The solids contents of sludge products are given in Table 5-4.

When sludge is land applied, it is considered to be a fertilizer substitute or a soil amendment. A "typical" dry ton of sludge will contain about 80 lb of available nitrogen (N), 50 lb of potassium (P_2O_5), and 20 lb of phosphorus (K_2O). The sludge can be applied to the land in a variety of forms — liquid, slurry, semisolid, or solid — with corresponding solids content from 2 to 95%. With a more liquid sludge (or sludge with lower solids content), the nutrient content may be very dilute. If a sludge contains 5% solids, each ton of slurry (about 250 gal) will contain 4 lb of available N, 2.5 lb of P_2O_5, and 1 lb of K_2O.

The economic feasibility of substituting sludge for fertilizer is a complex function of current fertilizer prices, sludge quality and processing costs, the distance that it must be transported to be reused, application and monitoring costs, long-term reimbursements and the cost of capital, the cost of alternative disposal, and the public support of the sludge recycling option — the cost of public education — which is in turn balanced against the cost of delay in implementing the sludge reuse program.

Sludge value = *f* (fertilizer prices, sludge quality, processing costs,
transportation costs, application costs, monitoring costs, cost of capital,
cost of public education, cost of alternative disposal and/or fines)

Each sludge product has a variety of associated costs. The capital expenditure for processing equipment and the operational costs of processing sludge are higher for pellet production, lime stabilization, and in-vessel composting than for windrow composting, reusing sludge cake, or simply liquid sludge. The revenue potential of a sludge product is linked with the fertilizer value of the product and the ease with which it can be applied and transported. Thus, sludge pellets have the highest revenue potential, with sludge compost, lime-stabilized product, cake, and liquid sludge providing progressively lower income potential. The transportation costs of finished

Table 5-4 Solids Content of Sludge Products

Sludge product	Solids content
Liquid sludge	1–3%
Thickened liquid sludge	≤10%
Dewatered sludge cake	18–31%
Sludge compost	40–60%
Lime-stabilized sludge	45–70%
Sludge pellets	90–98%

sludge product are associated with the water content of the product and the incorporation of additives; hence, sludge pellets are the least costly product to transport, while compost, cake, lime-stabilized sludge, and liquid sludge are progressively more expensive to distribute. Application costs follow the same pattern: sludge pellets can be broadcast without special equipment, as can sludge compost. Sludge cake and lime-stabilized sludge generally require a manure spreader, and applying liquid sludge requires the use of a specialized vehicle. Finally, monitoring costs are site specific and depend largely on the oversight of the state regulatory body. The sludge cost variables of the range of sludge products are given in Table 5-5.

5.5 SLUDGE MARKETS

There are a wide variety of markets available for sludge reuse. These can be generally classed as agricultural and nonagricultural uses. Agricultural uses include crops for human or animal consumption and pastureland. Nonagricultural uses include forest application, land reclamation, and a wide variety of green space enhancements such as parks, playing fields, airports, golf courses, nurseries, turfgrass production, and even landfill capping.

Options for liquid sludge reuse are limited by transportation costs and aesthetic impacts (unless injected), barriers that are progressively lower for sludge cake, lime-stabilized or composted sludge, and sludge pellets. Marketing options for sludge pellets are limited only by the level of effort a reuse program is willing to expend. The marketing possibilities are broad enough that it makes sense to define the characteristics of your "ideal" market before focusing on marketing strategies. Table 5-6 provides a range of sludge marketing variables.

Market diversity can be a method of increasing sludge reuse program reliability. Reuse programs with large quantities of sludge may conclude that having all of one's eggs in a single basket is an inherently unstable position and, hence, work to develop several markets for their sludge product. If there is a pressing need to maximize revenue, marketing sludge product in small, tasteful packages is likely to net the highest return. On the other hand, using sludge product to reclaim urban lots or strip-mined land is likely to net little return, but reuses large quantities of sludge for a public benefit.

If the POTW is located in a region with a long winter, there will be a strong seasonal component to local markets. If this problem cannot be resolved by building a large sludge storage area, it may make sense to develop a market with year-round demand, such as a bulking agent for commercial fertilizer. Some markets, such as

Table 5-5 Sludge Cost Variables

Sludge pellets	Compost	Lime Stabilized	Sludge Cake	Liquid sludge
High processing costs				Low processing costs
High capital expenditure				Low capital expenditure
High revenue potential				Low revenue potential
Low transportation costs				High transportation costs
Low application costs				High application costs
Low monitoring costs				High monitoring costs
Easily accepted by public				Difficult to win public acceptance

Table 5-6 Sludge Market Variables

Multiple markets	Single market
Maximize revenue	No revenue
Repetitive applications	One time use
Year-round demand	Seasonal market
Minimize liability	Maximize public contact
High public visibility	Low public awareness
Public benefit	Private benefit

final root zone cover for landfill closure, require large quantities of sludge product for a one-time use. Other markets, such as greenhouse potting mixture, use small quantities for an indefinite length of time.

Liability issues are related to the level of public exposure and the public acceptance of the sludge recycling program. Using sludge pellets as topdressing for urban ball fields may meet all federal exposure requirements, but public perception of the dangers of sewage sludge may create an unacceptable risk of litigation. Sludge application to forestland generally provides the lowest level of exposure to the public.

Out-of-sight, out-of-mind has been a password for municipal sludge disposal over the past few decades, perpetuating the myth that sludge products are unsuitable for beneficial reuse. If maintaining public acceptance and awareness of sludge recycling is an important goal, the municipality may choose to market its own sludge under a brand name and an easily identifiable logo. Maryland promotes its sludge recycling program by applying its sludge product to the White House lawn.

Table 5-7 provides an overview of the markets available for sludge reuse and the products that are appropriate for that market.

5.6 COSTING SLUDGE PROCESSING FACILITIES

The purpose of costing sludge processing facilities is twofold: to compare alternative sludge processing options and to budget for the construction and operation of a new sludge facility. Cost estimates of sludge processing operations are often incomplete.

The cost of constructing facilities is relatively straightforward, but sludge processing costs are more difficult to define. The cost of beneficial reuse should be defined as any additional costs beyond sludge digestion: the construction and the operation and maintenance (O&M) of sludge dewatering facilities, transportation, the capital and O&M costs of sludge stabilization, the costs of permitting and monitoring land application sites or the marketing costs for distributing the end product, the disposal costs of sludge processing by-products, and the product credit from sales. This cradle-to-grave costing of sludge processing options allows for an honest comparison of the different options. It is quite common for consultants to omit the costs of dewatering or transportation when comparing different sludge processing options, which skews the resulting cost estimates.

Sometimes the costs between different programs cannot be compared fairly. Programs which land-apply liquid sludge, dewatered cake, or lime-stabilized sludge are significantly different than programs that produce compost or pellets. Compost

Table 5-7 Matching Sludge Products and Markets

Sludge product	Sludge markets
Liquid sludge Sludge cake Sludge pellets	Silviculture State forestland Christmas tree farms Pine plantations Agriculture Pasture Cover crops Feed crops
Sludge cake Sludge compost Lime-stabilized sludge	Disturbed land reclamation Sand and gravel pit restoration Open sand areas Strip-mined land
Lime-stabilized sludge	Daily cover for landfills Acidic farmland
Sludge compost	One-time uses Final root zone cover for landfill closure Landfill reclamation Urban lot reclamation Construction of baseball fields Cemetery construction Installation of roadside curbing Landscape construction Repetitive uses Department of public works Parks department Turnpike authorities/highway departments Airports Private enterprises Cemeteries Golf courses and driving ranges Sod production Topsoil production Nursery potting mix Greenhouse potting mix
Sludge pellets	Bulk use Fertilizer bulking agent Pesticide/herbicide bulking agent Manufactured fertilizer base Bagged use Specialty fertilizer Potting soil component Nursery/landscape fertilizer Soil conditioner Turf fertilizer Trafficked lawn topdressing

and pellets can be marketed, and it is usually unnecessary to locate and permit the application sites. Programs that produce liquid sludge or sludge cake are generally required to locate and permit sites for land application, which may be a significant stumbling block to a reuse program. Some POTWs choose to purchase land to be used as an application site, which reduces the uncertainty involved in securing sites owned by others. A sludge reuse program that has made a significant investment in land application sites is difficult to compare to a reuse program that relies on privately or publically owned parcels of land: management costs are reduced, the program stability is increased, and the land investment may appreciate or depreciate, depending on the vagaries of the market. The costs of programs that have purchased their application sites cannot be compared directly with application programs that do not own sites.

The level of federal and state funding for sludge reuse projects varies across the country, from year to year, and from project to project. Grant money is often looked upon as "found money" that does not get tallied with the rest of the costs. The level of funding and the way the accounting is structured can impact cost estimates significantly.

Sludge processing costs are usually presented in dollars per dry ton, where the annual processing costs are divided by the dry tons of sludge. The pitfalls of this process emerge when the present values of processing options are calculated. Sludge quantities are usually projected to increase over time, O&M is subject to inflation, and the costs of construction are usually expressed as a loan paid off in a series of equal periodic payments. Present worth analyses tend to oversimplify the cost streams, producing inadequate estimates of sludge processing costs.

Comparing costs from different processing facilities is deceptive. The level of state or federal funding for a sludge processing facility has a big impact on the per dry ton processing costs, as does inflation. Construction costs vary across the country, and site-specific variables impact costs as well. Climate considerations and population density translate into cost factors. From one point of view, comparing costs from different sludge processing facilities is like looking at apples and oranges. For this reason, cost analyses other than your own should be taken with a grain of salt.

REFERENCES

1. Goldstein, N. "A Practical Approach to Facility Siting," *Biocycle* (February 1986).
2. "Memorandum of Understanding Between the City of Quincy and the Massachusetts Water Resources Authority," 1988.
3. "Sludge Management: Two Decades of Progress," Sludge Management Program, Metro, Municipality of Seattle, Washington (August 1986).
4. O'Hare, M., L. Bacow, and D. Sanderson, *Facility Siting and Public Opposition* (New York: Van Nostrand Reinhold Co., Inc., 1985).

6 LAND APPLICATION OF LIQUID SLUDGE

6.1 INTRODUCTION

The land application of liquid sludge has a venerable history and has been used for centuries to improve crop yields. Sludge was land applied in the U.S. for generations before the advent of cheap, plentiful chemical fertilizers in the 1950s, when the practice became virtually obsolete. The present trend in wastewater treatment toward the land application of sludge is not a new development, but the revival of an old solution.

The land application of liquid sludge conforms to the U.S. EPA policy of beneficial reuse of an environmentally acceptable product. Beneficial reuse lowers the net cost to society by diverting sludge from landfills or incinerators. Liquid sludge application provides a method of reusing sludge at minimal cost, reducing the taxpayer burden for sludge management. Farmers that land apply sludge reduce their costs for inorganic fertilizer while adding organic matter to the soil. Even though the land application of liquid sludge benefits both taxpayers and farmers, there are inherent problems to liquid sludge application.

Rural farms may be so far from an urban POTW that the cost of transporting liquid sludge is prohibitive. Suburbia has spread through much of the farmland that formerly surrounded our cities, and farming practices must now accommodate neighbors with suburban noses. Programs must be aesthetically acceptable to local residents and landowners to be successful. The volume of liquid sludge produced at medium-sized and large facilities can be daunting. Sludge production is constant throughout the year, but the availability of land for disposal may not be. Sludge is typically stored during periods when crops are on the fields or during foul weather and poor ground conditions. It is difficult to obtain long-term sites for liquid sludge application programs. POTWs that are committed to a program of land application often purchase the land that they use as application sites to reduce program risk by ensuring that a site will be available when they need it. This can substantially raise the cost of a liquid application program.

A large quantity of data must be collected and analyzed to satisfy local, state, and federal regulatory authorities as well as to meet program management requirements. Data on sludge, soil, crop tissue, application rates, and groundwater must be reviewed regularly to ensure that the program remains environmentally safe. Coordinating local, state, and federal regulations, sludge handling considerations, and the schedules of local farmers involves a significant level of effort.

Sludge application rates are usually limited by the nutrient loading required by the crop rather than the levels of sludge contaminants. If the levels of heavy metals in sludge are high, then the allowable application rates may be so low that sludge nutrients are not enough to reduce commercial fertilizer application. The impact of heavy metals in sludge applied to cropland can be reduced by growing crops that do not accumulate heavy metals. In general, leafy vegetative tissues accumulate higher levels of heavy metals than grain crops.

The land application of liquid sludge offers a low-cost and technically uncomplicated method of reusing the material. If a city produces clean sludge and an application site can be located and permitted, liquid application can be an inexpensive method of beneficially reusing wastewater residuals.

6.2 AGRICULTURAL APPLICATION

Sludge is applied to agricultural land as a substitute for commercial chemical fertilizers. The primary difference between sludge and chemical fertilizer is that the volatile solids in sludge typically total more than one half of the total solids, making sludge a good source of organic matter. It has lower levels of nitrogen, phosphorus, and potash than chemical fertilizers; these nutrients are associated with the organic fraction of the sludge, which means that the nutrients are unavailable for plant uptake until the microorganisms in the soil can break down the organic matter.

Sludge nutrients are in a water-insoluble or "slow release" form. Synthetic fertilizers are typically water soluble, so the applied fertilizer is readily accessible to plants. If there are more nutrients in an application of synthetic fertilizer than the plant requires at that particular time, nitrogen and other nutrients may be lost through leaching, runoff, or soil erosion. The surface runoff from heavily fertilized fields can be a significant source of nitrogen and phosphorus in nearby streams, fostering the growth of algae, which use up the oxygen in water and can virtually suffocate aquatic life at night. Since nutrients are in much lower concentrations in sludge than in chemical fertilizers, this type of nutrient pollution is not commonly associated with sludge application. A program which uses sludge instead of synthetic fertilizers can actually help prevent groundwater degradation.[1]

Chemical fertilizers can burn plants through excessive nutrient loadings and can kill the bacteria, worms, and other creatures that are essential for healthy soil. Sewage sludge, with its organically bound nutrients and large fraction of organic matter, builds soil integrity and feeds plants over a longer period of time.

6.2.1 Effects of Sludge on Cropland

Amending cropland with sewage sludge increases crop yields substantially. The nitrogen and phosphorus contents are only part of the story; each of these elements can be added by applying inorganic fertilizer. Adding the organic matter in sludge to the soil has a number of benefits. The plant nutrient retention capability or cation exchange capacity (CEC) of the soil is increased, the water holding capacity of the soil is raised, the physical properties of the soil relative to tillage are improved, and soil loss from wind or water erosion is reduced. The organic matter in sewage sludge reduces the bulk density of soils while increasing their aggregate stability, water retention, and hydraulic conductivity. In sandy soil, the organic material in sludge fertilizer acts like moss in a sieve; in clay soil, where plants can become suffocated or waterlogged, the organic material lightens the soil and allows air to reach the roots.

The soil conditioning properties of sludge have exceptional benefits in revegetating disturbed land. In areas where subsoil has been left on the surface without a layer of topsoil, plants are unable to grow and the land cannot revegetate. To upgrade subsoil to topsoil, organic matter must be added. Disturbed land cannot be successfully revegetated through the application of commercial fertilizer alone. Sludge, with its robust organic content, is well suited to the task. Whether it is to revegetate strip-mined land or urban lots, create a final root zone cover for landfill closure, or restore land disturbed through urban construction, sludge application is a successful solution to a range of environmental problems.

One of the greatest concerns of reusing sludge is that the pathogenic organisms associated with human waste will contaminate the area to which it is applied. A recent EPA study found that pathogen levels in processed sludge products are surprisingly low.[2] Not only is the level of pathogens low, but the pathogenic organisms associated with sludge products appear to be short-lived in the soil. Studies have found that after 1 year the density of pathogenic organisms in sludge-amended soil is essentially the same as for soil with no sludge application. The potential hazards to human health from sludge-related microorganisms are negligible.

6.2.1.1 Soil pH

The greatest area of concern is that trace elements will accumulate in the soil and be accumulated by plants. In the past, pH has been considered to be the major limiting factor controlling the migration of metals in the soil. The depredations of acid rain are caused largely by the release of metals that had been previously bound to the soil. When a lowered soil pH allows these molecules to migrate into the food chain, plants and animals may receive toxic doses of indigenous metals. The solubility of these elements is inversely related to pH, except in the case of molybdenum and selenium. In the laboratory, metal ions are soluble in a soil solution to a pH of 7.0. At a higher pH, most metals form a sparingly soluble metal-hydroxide complex which precipitates from the soil solution.[3] The formation of hydroxides, carbonates, and phosphates and complexation and chelation with other anions will immobilize the trace elements found in sludge. As the pH drops, the metals are more apt to migrate. Recently, however, it has been found that the direct pH-metals retention relationship is heavily influenced by the amount of organic material in the soil. For this reason, the land application of sludge may serve to enhance the ability of soils to retain their metals by increasing the amount of organic matter in the soil.

Most states require that sludge be applied to soils with a pH adjusted to 6.5 or higher to preclude the migration of metals from the zone of sludge incorporation. The EPA agrees that soil pH is one of the strongest influences on the capability of plants to absorb pollutants from the soil/sludge matrix. However, in some cases the data from low-pH studies were also used in the exposure assessment model to develop numerical limitations for agricultural practices. For this reason, the EPA concludes that its numerical limits protect a majority of U.S. soil conditions by requiring pH control for all agricultural land practices regulated under the Part 503 rules.

Heavy metals in sludge-amended soils show limited downward movement through the soil and are usually retained in the top 0 to 15 cm. When sludge of low metals concentration is land applied, most of the metals interact with the soil to form insoluble compounds and complexes. The uptake of metals by plants is very low in comparison to the initial sludge concentrations.

In some soils, sludge loadings of heavy metals are plant micronutrients rather than contaminants. Zinc and copper are often deficient or becoming deficient in soils that have been used as cropland for decades, and sludge application is preferable to applying mineral salts. In iron-deficient calcareous soils, sludge is a remarkably beneficial iron fertilizer that cannot be replaced with any normal commercial fertilizer.

6.2.2 Research on Crop Response

Sewage sludge is a good source of plant micronutrients. The two nutrients of greatest economic importance are nitrogen and phosphorus. The multiplicity of crops that need these nutrients and the site-specific nature of farming have given rise to a broad body of research on crop response to sludge loading rates.

Research on crop response to sludge application expands the base of knowledge on the effects of sludge application, and it improves the credibility of a sludge management program by demonstrating the effects of the local sludge on area soils. Sludge research projects are concrete demonstration projects, and most management programs are involved in some level of sludge research. Areas of particular interest include loading rates, the effects of sludge on plant concentrations of heavy metals, crop yields, and the protein content of the grain.

Research showed that grain yields from wheat cultivars grown with dried sewage sludge were comparable to the yields obtained with recommended rates of inorganic fertilizer. The yield per acre of corn and hay grown on sandy loam amended with sewage sludge was higher than the yield from plots fertilized with ammonium nitrate. Since the nitrogen supplied was the same in both cases, researchers concluded that the increase in yield was attributable to the increased supply of potassium and other micronutrients and the residual nitrogen. Maximum nitrogen uptake and hay yield from tall fescue were obtained with combinations of sewage sludge and inorganic fertilizers, with the sewage sludge providing needed nutrients even after the nitrogen from inorganic fertilizers was depleted.

Protein yields of grain that is grown in sludge-amended soils are typically increased by up to 20%. Growers are paid protein premiums for grain with higher protein content, so sludge application can result in increased yield coupled with increased crop value and reduced fertilizer costs.[4]

A typical example of sewage sludge research was conducted by the University of Arizona in Tucson from 1987 to 1988. This project studied the effects of liquid sewage loading rates on the vegetative growth, yield, and quality of barley grain and straw. The plants were grown in pots filled with sandy loam; the anaerobically digested liquid sludge was from the Ina Road Sewage Treatment Plant near Tucson. The sewage sludge had a pH of 7.6, 1.5% total solids, 9% total nitrogen, 5.1% phosphoric acid, and 0.4% potash on a dry weight basis.[5]

The crop loading rates consisted of seven treatments, each replicated four times. The yields and the concentrations of heavy metals are given in Table 6-1. This experiment shows that yields of straw and grain from plants grown in sludge-amended soils are higher than for plants grown without fertilizer or with inorganic fertilizer. Plant and soil concentrations of copper and zinc increased substantially, while grain and soil concentrations of nickel rose slightly. Cadmium and lead concentrations were unaffected by the application of sludge.[6-8]

6.2.3 Liquid Application in Southwestern Ohio

When the North Regional Wastewater Treatment Plant in southwestern Ohio began a land application program for liquid sludge, one of the project goals was to demonstrate that the liquid injection of sludge on agricultural land could benefit and be coordinated with a successful agricultural operation. Once the decision had been made to define the sludge management program as a demonstration project, their first step was to acquire a base amount of land that would allow research to be carried out. When the research was completed, the program would be able to secure contracts with private landowners for the additional land required.

By owning their application site the land would always be available for sludge disposal, even if a crop had to be destroyed in an emergency situation. Direct ownership of the land increased program flexibility and reduced the risk of ending up without an application site. In the early stages of the project the treatment plant acquired and leased three parcels totalling 320 ha or 790 acres. A "permit to install" application for sludge disposal was processed by the Ohio EPA, and public meetings were held with local citizens to explain the details of the proposed plan.

Once permitted, major site improvements were carried out to increase the efficiency of the sludge application program. Plastic tiles were added to wet areas so these areas could be used, several acres were cleared of tree growth, tree-lined fencerows were removed to make large fields from several small ones, and an equipment shed was constructed to protect the sludge application equipment when not in use.

The land was leased to a local farmer to perform the agricultural operations. The sludge management program priorities are to dispose of sludge at the proper agronomic rates on a schedule that suits the sludge production and storage capability. The farmer's priorities are planting and harvesting crops for a minimum cost and maximum yield. With the help of good communication most of these criteria have been met.

The land is not available for planting until the sludge application is completed, and general provisions for crop selection are defined to ensure the proper uptake of nutrients. Fertilizer can be used only if the combination of sludge and fertilizer does not exceed recommended agronomic requirements for the crop planted. Each year the farmer must report all herbicide, fertilizer, and other chemicals used for each field, as well as the crop yield.

To be compatible with agricultural operations, the sludge must be stored during the growing season when crops are in the ground. Since Ohio winters are normally cold and wet, storage must be provided during this period as well because direct

Table 6-1 Crop Yields and Metal Concentrations in Sludge-Fertilized Barley — Greenhouse Results from the University of Arizona at Tucson

Material sampled	Treatment	Yield (g/pot)[a]	Cadmium (μg/g)	Copper (μg/g)	Lead (μg/g)	Nickel (μg/g)	Zinc (μg/g)
Sludge	—	—	10	886	218	53	118
Grain	1	28	0.3	10	5	32	22
	2	35	0.4	9	1	33	22
	3	40	0.3	12	2	38	30
	4	46	0.4	12	3	36	32
	5	46	0.4	13	4	35	35
	6	41	0.4	12	1	35	40
	7	43	0.4	16	6	40	40
Straw	1	54	0.5	7	9	54	30
	2	61	0.4	7	9	38	37
	3	71	0.5	8	10	47	48
	4	82	0.6	7	12	44	52
	5	79	0.7	7	17	50	46
	6	90	0.5	10	10	51	36
	7	92	0.4	10	4	47	45
Soil	1	—	1.3	13	20	58	79
	2	—	1.6	15	22	64	78
	3	—	1.4	15	20	65	87
	4	—	1.5	15	16	57	78
	5	—	1.6	17	29	67	82
	6	—	1.2	20	19	70	96
	7	—	1.4	27	16	62	86

Source: Day, Solomon, Ottman, and Taylor.[5]
Note: Treatments:
1. Control (a soil with no fertilizer applied)
2. Inorganic ammonium nitrate applied in recommended rates for barley (112 kg/ha)
3. Liquid sewage sludge to provide the recommended rate of plant-available nitrogen
4–7. Liquid sewage sludge rates to provide plant-available nitrogen in amounts equal to 2,3,4, and 5 times the recommended plant available nitrogen, respectively

[a] Plants were grown in pots of sandy loam.

injection into frozen or wet ground is impossible. It was determined that 120 days of sludge production design flow would be adequate sludge storage. Storage was provided at the disposal sites rather than the treatment plant to allow sludge transport to take place during off-season months so all efforts could be concentrated on sludge application during the peak application periods. Above-ground tanks were found to be the most cost effective to construct and operate, and a total off-site storage capacity of 4.95 million gal was installed.

Subsurface injection of liquid sludge was chosen as the preferred technique of sludge application because it provides better odor control, maximum fertilizer value, improved aesthetics, and decreased potential for the public to come in contact with the sludge. The sludge is incorporated directly into the soil and is never seen on the surface of the field. There is a wide range of equipment available for sludge injection. Injecting sludge 12 to 18 in. below the surface is very hard on equipment, and wet or rocky conditions may make it impossible. Both articulated tanker trucks and tractor-pulled tank trailers are used for liquid injection, although the latter provide better handling and larger capacity.

The program purchased both a 320-hp articulated tank truck with a capacity of 3200 gal and a 290-hp four-wheel-drive tractor with a 5000-gal tank trailer. Operators prefer to use the tractor-pulled tank trailer because of its maneuverability and higher capacity. In addition to the two sludge injection rigs, an irrigation system with a traveling gun, aluminum irrigation piping, and a high-pressure portable pump was purchased to spray sludge a distance of 150 ft. Although sludge injection remains the method of choice, the irrigation system is used during periods of wet and frozen ground conditions when the injection equipment cannot be used. The sludge irrigation system also can be used when crops are on the field, as long as it is applied more than 60 days before harvest.

Data are collected to address two areas of concern: nutrient loading from the crop production and heavy metals accumulations in the soil from sludge application. Each time sludge is applied, the amount of metals added to the specific field is recorded and added to the previous accumulations to maintain a lifetime record of metals application to the field. Soil and plant tissue analyses are also recorded to develop a history of the fate of various parameters, particularly heavy metals.

As a part of the sludge management program, 18 monitoring wells and 3 springs are used to monitor any potential effects of sludge disposal on groundwater. The wells and springs are sampled quarterly. Field measurements such as temperature, conductivity, alkalinity, and pH are taken on site, while chlorides, dissolved solids, fecal coliforms, ammonia, nitrite, nitrate, total organic carbon, and metals are determined in the laboratory. To date there has been no change in any of the parameters monitored, indicating that local water supplies are unaffected by the sludge application program.

The North Regional WWTP program is manageable and cost effective. By purchasing its own sites for land application, the program has removed one of the major obstacles to liquid application. Storage issues and material transport considerations were rigorously addressed and resolved up front, and the range of sludge application vehicles and technology provides flexibility in application methods during the growing season.[9]

6.3 FOREST APPLICATION

In areas where forestland is more common than farmland, silviculture is beginning to replace agriculture as a market for liquid sludge. Forests have fewer residents than farmland and most forest products are not food chain crops, so public health concerns and land application regulations are generally less restrictive for forest application sites than they are with agricultural crops.

Forests occupy about 40% of the land in the contiguous U.S. and are typically located in well-drained sites that do not experience the periodic flooding of alluvial agriculture areas. With years of humus collecting beneath the trees, sylvan soils are generally porous, minimizing the surface runoff of applied nutrients. The humus in forest soils also contains organic carbon to immobilize the available nitrogen and metals in sludge. Unlike many agricultural crops, forests have a perennial root system to allow year-round uptake of available nutrients. Forest soils are often nutritionally impoverished, so sludge application can substantially increase the nutrient levels in the soil as well as the organic content. Nitrogen and phosphorus are the limiting nutrients for many U.S. forests, and both are found in municipal sludge. Finally, forest sites are generally located away from population centers and are used for dispersed recreational activities, minimizing the level of human contact with recently applied sludge.[10]

The economic effectiveness of forest sludge applications depends greatly on accelerated growth and the effect on wood quality. One of the drawbacks of reusing sludge in silviculture projects is that trees take so long to mature that the return on investment is very slow. It is relatively inexpensive to apply liquid sludge to forestland, but it may be a half century before the economic benefits of increased tree growth can be realized. Although some work on the long-term effects of sludge on growth rates and wood quality has been completed, the results in this area are still preliminary.[11]

6.3.1 Effects of Sludge on Forestland

Increased tree growth is a nearly universal effect of sludge application in the forest. For years following a sludge application, major portions of the supplemental nutrients and trace elements are retained as unavailable, undecomposed forms in the humus of the forest floor. Nutrients in available forms moving from the forest floor are readily taken up by plants, with no significant fertility changes in the surface or subsurface soils. This means that a single sludge application can provide a slow, steady supply of nutrients that lasts for many years.

The plant species distribution is not affected by sludge application, but the quantity and vertical distribution of cover beneficial to important wildlife species are substantially improved.[12] The sludge-applied nitrogen and phosphorus is rapidly taken up by trees, increasing foliar nutrient concentrations.

While native plants are accustomed to the low ambient nutrient levels of forest soils, they respond to sludge application with significant increases in biomass and nutrients. Annual herbaceous growth increased 200% 1 year following sludge application and remained 50% higher than untreated areas 3 years later. The undergrowth becomes lush and thick, with the greatest gains in understory vegetation in the lower 6 ft, where the greenery is well within reach of browsing animals. Protein and phosphorus may increase 20 to 50% in important wildlife forage plants within 1 year of sludge application and persist for 3 years following application. Elk and deer flock to the sludge-treated woodlands and grow sleek from browsing on the denser, enriched forage. For several years after the sludge is applied the deer produce more twins in the spring, taking biological advantage of the increased food supply.[13]

6.3.2 Research on Tree Response

Research in application rates, methods, and results has shown that sludge application increases tree growth dramatically. Figure 6-1 is a cross section of a tree that had a single application of sludge 12 years after it was planted. It is clear that the tree experienced accelerated growth for many years after a single application of liquid sludge. The annual diameter increase of trees receiving only one application of sludge can be 50 to 400% higher than unfertilized trees. Table 6-2 provides data on the growth response of trees after sludge application.

The accelerated wood growth does affect the density of the harvested trees; 10 to 15% lower specific gravity has been noted for sludge-grown wood. This apparently

Figure 6-1 Cross section of a tree with a single sludge application at year 12. (Courtesy of MWRA.[20])

Table 6-2 Growth Response for a Young Plantation after Application of 160 Dry Tons per Hectare of Sludge

Species	Current age (years)	Increase over control (%) Height response	Diameter response
Douglas fir	4	64	150
Western hemlock	4	370	695
Sitka spruce	4	140	225
Western red cedar	4	65	115
Cottonwood	4	320	585
Hybrid cottonwood	4	590	895
Poplar	4	1190	—
Redwood	3	575	1035
Grand fir	3	200	1250
Sequoia	3	285	1035
Basswood	3	200	125
Birch	3	900	690
All conifers		270	645
All deciduous		640	560

Source: Cole, Henry, Schiess, and Zasoski.[10]

reflects the change in forest site quality produced by the sludge treatment and is within the specific gravity range found in trees on higher site land.[14]

Sludge application is most effective on stands where the trees are well established. When sludge is applied to trees that are a minimum of 6 ft high, well above deer browsing range, the increased productivity that follows sludge application is reflected in the tree seedlings rather than the undergrowth. Not only should the stand of trees be well established; it appears that younger stands of trees — from 10 to 30 years old — are more responsive to sludge application.

Sludge application to the canopy of a young forest may damage the foliage. Applications should be limited to periods of high rainfall or to the dormant stage of the growth cycle. In well-established forests, sludge can be sprayed beneath the canopy with no seasonal limitation.

Research has shown that it is impractical to apply sludge to cleared areas prior to planting.[15] The undergrowth responds so well to sludge application that the tree seedlings can't effectively compete with the plants for moisture or space. Unless the undergrowth is controlled, the tree seedlings grow poorly and have a high rate of mortality. The lush undergrowth in sludge-treated areas has a relatively high nutritional value, and deer may selectively browse the site and further increase seedling mortality. Thick undergrowth in the fertilized habitat can also provide excellent cover for voles, which, if left uncontrolled, will girdle young trees. Trees grow so poorly in clear-cut areas that Seattle has experimented with using sludge application as a method of vegetation control in areas such as powerline right-of-ways where trees are unwelcome.

Food chain studies are used to assess the effects of bioaccumulation of sludge contaminants. Bioassays of plant and animal tissues from organisms exposed to sludge-fertilized soils in upland forest and laboratory trials show minor accumulations of toxicants to levels that would not be harmful to forage plants, herbivores, or carnivores at higher trophic levels, including humans. Laboratory food chain studies of sludge-soil-grass-mice-raptors (hawks and owls) and sludge-soil-earthworms-woodcocks yielded minor accumulations of cadmium, nickel, and chromium in mice liver and kidney tissues. These accumulations are similar to those found in free-ranging small mammals sampled in agricultural settings which have had sludge applications.[16]

Residence time and mobility of fecal coliform and total coliform bacteria have been studied in the Pack Forest in Washington State, and other pathogenic organisms have also been studied to some extent. In a 10-cm application of sludge, the residence time of fecal coliforms is less than 2 years, and less if the sludge is applied to recently harvested sites or is applied during the summer. Fecal coliforms do not appear to migrate down into the soil or the groundwater table.

The limiting factor in forest application of sludge is typically leaching of nitrates into the groundwater, and application rates are designed to prevent overloading the forest ecosystem with nitrogen. Most of the nitrogen in sludge is in organic and ammonium forms and, hence, is not easily leached. However, following mineralization and nitrification it can be easily leached. The rate of leaching is controlled by a series of processes that includes mineralization, nitrification, denitrification, volatilization, immobilization by soil organisms, and vegetative uptake. The complexity of contributing factors means that nitrate leaching must be determined empirically. Studies at Pack Forest have shown that a single application of sludge in excess of 40 dry tons per hectare could result in some nitrate leaching during the first year of application.[10] Heavy rainstorms after sludge application also may cause surface erosion if the sludge has not had enough time to stabilize.

6.3.3 Forest Application in Seattle, Washington

In 1970, the Seattle wastewater treatment district, (Metro), decided to stop dumping its sludge into Puget Sound. In a state where population density is low and forestland is common, sludge application for silviculture is an attractive option. Metro made its first application of digested sludge to forestland in 1973. The sites were the Pope and Talbot Co. forest plots in Jefferson County and the University of Washington's Pack Forest near Eatonville. The projects had three goals: evaluating timber response to sludge application rates, monitoring environmental affects, and testing the feasibility of a water cannon mounted on a tank truck to spray the sludge. The Pack Forest project, which began as a research effort in reforestation, expanded into a 10-year project involving a major amount of Metro's sludge on a 200-acre site. The silviculture experiment was such an unbridled success that by 1987 Metro had purchased a total of 4489 acres of land to ensure a permanent site for sludge-aided silviculture.[17]

The technology developed to apply sludge to agricultural sites is not easily transferable to forest sites. Liquid injection is not feasible in wooded areas, and forest application vehicles must be able to negotiate rudimentary roads with sloped and broken terrain. In the 1970s Seattle had to develop its own vehicle to apply liquid sludge, but since then the Ag-Chem Equipment Company of Minneapolis, MN has developed a vehicle that can be used to distribute liquid sludge through a mature forest. This sludge-slinger has a 2000-gal-capacity tanker truck and a dual supercharged pump system that allows the use of a nozzle. Liquid sludge is sprayed from the vehicle 130 ft into the forest at a rate of 375 gal/min.

To reduce transportation costs, the sludge is centrifuged at Metro's West Point Treatment Plant. Sludge cake is loaded onto a transport truck-trailer system which holds 28 to 30 tons, and the cake is transported to an application site. At the site, the dewatered sludge is mixed with water in an in-ground mixing tank and stored in an on-site portable storage tank. The application vehicle loads liquid sludge from either the storage tank or a mobile transfer trailer, which acts as a pipeline on wheels between the storage tank and the application truck. The total application of sludge per acre is nearly 35,000 gal, requiring 19 applicator vehicle loads.

Initial research in the mid-1970s used relatively heavy sludge applications to newly clear-cut sites. At the time sludge management was governed by the "disposal philosophy", which favored maximum application to a minimum land base. Heavy applications also made it easier to identify the public health issues. It was not known what application rates produced optimal tree growth, and the technology for forest application was as yet undeveloped.

Early research clearly demonstrated dramatic growth responses, but also identified problems associated with heavy applications. As a result of the surface runoff and leaching of sludge during the early years of Seattle Metro's silviculture projects, application rates have been substantially reduced since the 1970s. Nitrogen leaching in particular led researchers to experiment with application rates. The sludge is now applied to the forest in five separate layers with a month or more between each application to allow each layer to stabilize. The sludge application is repeated every 5 years.[18]

Sludge application is not a pretty sight. The tanker truck lurches down the skid trail to the application site, where it stops, points the nozzle, and sprays sludge through the trees. The forest understory is covered with a fine film of liquid sludge. Most insects and spiders are killed or driven out of the area and may not reestablish themselves for up to 1 year. Rodents and birds temporarily leave the sludge application area to return in less than 1 week.

Public acceptance is one of the key issues for success in implementing a land application program. The relative newness of forest application when compared to agricultural application makes public education a critical issue for silviculture operations. Several technically sound silviculture projects proposed in Washington did not receive public support and subsequently died. With this in mind, Metro made public information a major element in its forest application program. Metro conducts site tours for the public, distributes an information newsletter on its silviculture operations, and gives presentations to civic groups, city councils, and professional societies. By addressing public concerns through research and education, Seattle Metro has been able to establish a highly successful sludge reuse program.[19]

REFERENCES

1. Cheney, R. "Twenty Years of Land Application Research," *Biocycle* (September 1990).
2. Yanko, W.A. "Occurrence of Pathogens in Distribution and Marketing of Municipal Sludges," EPA/600/1-87/014 (1988).
3. Hahne, H.C.H. and W. Kroontje. "Significance of pH and Chloride Concentration on the Behavior of Heavy Metal Pollutants: Mercury, Cadmium, Zinc and Lead," *J. Environ. Qual.* 2:444–450 (1973).
4. Utschig, J.M., K.A. Barbarick, D.G. Westfall, R.H. Follett, and T.M. McBride. "Liquid Sludge vs. Nitrogen Fertilizer," *Biocycle* (August 1986).
5. Day, A.D., M.A. Solomon, M.J. Ottman, and B.B. Taylor. "Crop Response to Sludge Loading Rates," *Biocycle* (August 1989).
6. Annis, C. "From Treatment Plant to Farm Fields," *Biocycle* (April 1990).
7. Goldstein, N. "Land Application in the Southwest Desert," *Biocycle* (March 1989).
8. Barnes, D. "Metrogro Sludge Program Boosts Farm Profits," *Biocycle* (February 1989).
9. Royer, L.S. and N. Brookhart. "Common-Sense Liquid Sludge Management," *Water Environ. & Technol.* (December 1990).
10. Cole, D.W., C.L. Henry, P. Schiess, and R.J. Zasoski. "The Role of Forests in Sludge and Wastewater Utilization Programs," Proceedings of the 1983 Workshop on Utilization of Municipal Wastewater and Sludge on Land, Page, Gleason, Smich, Ishkander, and Sommers, Eds. Sponsored by the U.S. EPA and the University of California, Riverside (1983).
11. Brockway, D.G. "Forest Application of Municipal Sludge," *Biocycle* (October 1988).
12. Haufler, J.B. and D.K. Woodyard. "Influences on Wildlife Populations of the Application of Sewage Sludge to Upland Forest Types. Final Project Report," Department of Fisheries and Wildlife, Michigan State University, East Lansing (1986).
13. Campa, H., D.K. Woodyard, and J.B. Haufler. "Deer and Elk Use of Forage Treated with Municipal Sewage Sludge," in *The Forest Alternative for Treatment and Utilization of Municipal and Industrial Wastes,* D.W. Cole, C.L. Henry, and W.L. Nutter, Eds. (Seattle: University of Washington Press, 1986), pp. 181–198.
14. Brockway, D.G. and P.V. Nyugen. "Municipal Sludge Application in Forests of Northern Michigan, a Case Study," in *The Forest Alternative for Treatment and Utilization of Municipal and Industrial Wastes,* D.W. Cole, C.L. Henry, and W.L. Nutter, Eds. (Seattle: University of Washington Press, 1986).
15. Wilbert, M. and S.G. Archie. "Management of Sludge-Treated Plantations," *Municipal Sludge Application to Pacific Northwest Forestlands,* C.S. Bledsoe, Ed., Institute of Forest Resources, Contrib. No. 41, College of Forest Resources, University of Washington, Seattle (1981), pp. 101–103.

16. Brockway, D.G. et al. "Municipal Sludge Application" (1986).
17. Henry, C.L. and D.W. Cole. "Pack Forest Sludge Demonstration Program History and Current Activities," in *The Forest Alternative for Treatment and Utilization of Municipal and Industrial Wastes,* D.W. Cole, C.L. Henry, and W.L. Nutter, Eds. (Seattle: University of Washington Press, 1986).
18. "Sludge Management: Two Decades of Progress," Sludge Management Program, Metro, Municipality of Seattle, Washington (August 1986).
19. Machno, P.S. "Seattle's Diversified Approach to Sludge Use," *Biocycle* (July 1986).
20. MWRA Residuals Management Team, "Surveying the Nation's Sludge," *The Inside Scoop* Vol. 2, No. 1, Spring 1991.

7 SLUDGE DEWATERING

7.1 INTRODUCTION

Dewatering may be the only processing required before digested sludge can be beneficially reused. The sludge cake produced by common dewatering techniques has a consistency similar to turkey dressing. A nonfluid material, it is easily handled and surprisingly nonoffensive, and it can be readily land applied by conventional manure spreaders. It is not technically difficult to spread dewatered sludge cake, but locating and permitting acceptable application sites may be problematic. Assuming that application sites can be secured, the dewatering process removes more water from the sludge than gravity thickening, with a greater volume reduction. With less volume, the capital and operating costs of land applying the sludge or subsequent processing are reduced. Dewatering sludge has an enormous impact on sludge disposal costs: when sludge is dewatered from a concentration of 2% solids to 20% solids, the volume of sludge is reduced by 90%. As shown in Figure 7-1, if 100 gal of sludge at 2% solids is dewatered to 32% solids, the volume of sludge to be disposed of is reduced to 6.25 gal — not a bad way to cut transportation costs!

Water in the sludge particle exists in four major phases: free water, capillary water, colloidal water, and intracellular water, as shown in Figure 7-2. Free water can be easily separated from sludge by gravity. Capillary and colloidal water can be removed, usually after chemical conditioning, by mechanical methods including centrifuges, belt presses, and vacuum filters. Intracellular water can be separated from the sludge particle by breaking the cell structure with thermal treatment. This type of dewatering is discussed in Chapter 10.

7.2 NATURAL DEWATERING METHODS

Dewatering technologies can be categorized as natural or mechanical methods. Natural dewatering systems include sludge lagoons, sand drying beds, Wedgewater drying beds, *Phragmites* reed beds, and dewatering via freezing. These dewatering methods require far less purchased power than mechanical thickening and usually rely on the force of gravity, solar power, or biological processes as the source of energy for dewatering. Natural dewatering systems are usually less expensive than mechanical dewatering systems and are typically less controllable.

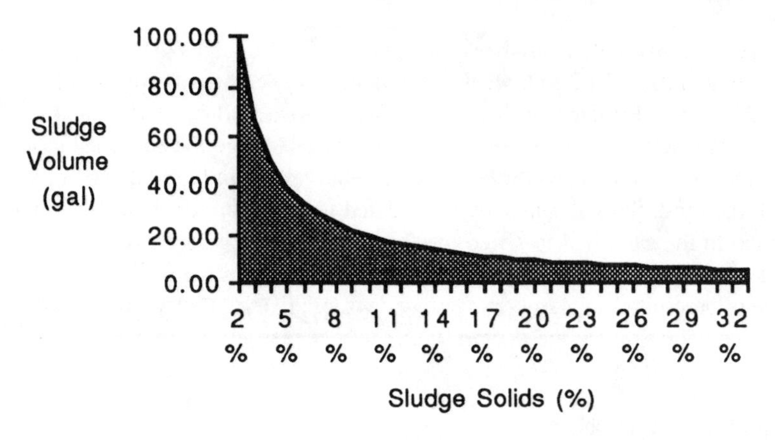

Figure 7-1 Sludge volume decreases as sludge solids increase.

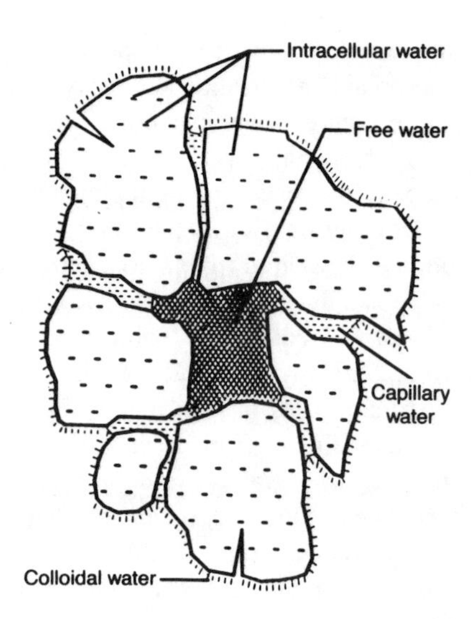

Figure 7-2 Water distribution in a sludge particle. Digested sludge has 5% total solids. (From Girovich, M. J., *Water Eng. Manage. Mag.,* Mar. 1990. With permission.)

7.2.1 Sand Beds

For small- to moderate-scale POTWs, sand drying beds have been the traditional method of sludge dewatering. Under favorable conditions they can produce a sludge cake comparable to cake from mechanical dewatering methods. Sand drying beds need little operator attention, but are usually restricted to digested sludge because raw sludge is smelly, attracts insects, and does not dry well when applied at reasonable depths. The oil and grease associated with raw sludge clog the sand bed and impede drainage.

Sand drying beds dewater the sludge by gravity drainage, by capillary action drawing the water through the sand, and by evaporation from the surface of the bed.

The water drains rapidly from the sludge in the first 1 to 3 days. After this is a period of slow evaporation for 2 to 6 weeks, producing a cake of up to 40% solids in well-digested sludges. Primary sludge dries faster than secondary sludge, and digested sludge dries faster than raw sludge. In well-digested sludge the entrained gases tend to float the sludge solids and leave a layer of relatively clear liquid that readily drains through the sand. Sludge is usually flocculated with a polymer conditioner before it is applied to the sand bed to speed the separation of sludge solids from the water.[2]

To a certain extent, sand drying beds are at the mercy of the weather. The amount and rate of precipitation, number of sunny days, relative humidity, and wind velocity all have an effect on sludge drying. If the weather is below freezing the entrained water will be immobilized on the sand drying bed until there is a thaw.

Drying beds are usually constructed of 10 to 30 cm of sand laid over 20 to 50 cm of gravel. The sand is typically 0.3 to 1.2 mm while the gravel is graded up from 2 cm. The water percolates through the gravel, and the sand is laid on top to filter out the finer particles of the liquid sludge. Underdrain piping has a minimum diameter of 10 cm and a minimum slope of about 1%. Collected filtrate is returned to the head of the treatment plant. Sand beds are occasionally enclosed to protect the drying sludge from inclement weather and to contain odor and insects. Experience has shown that an enclosed bed uses only 67 to 75% of the area of an open bed. Good ventilation is important to control humidity and optimize evaporation.[2]

Removing the dewatered sludge from sand drying beds can be a problem. To provide permeability for rapid dewatering, the sand layer should consist of uniformly graded particles with less than 5% fines. In a typical soil mix the fines provide structural stability, and the lack of fines or range of particle sizes decreases the ability of the soil to support loads. Methods of removing the dewatered sludge from a sand bed include hand labor and various forms of mechanical devices that are not directly supported on the sand. Small tractors or loaders cannot operate on the loose sand of a conventional sand drying bed because the lack of fines results in shift or shear under load conditions. Mechanical systems have had mixed success because of their high capital and operating costs as well as maintenance problems.[3]

Concrete slabs that accommodate the tire width of a tractor can be built into the sand drying bed, facilitating tractor operation and reducing the effective surface area of the sand filter. Geotextiles designed to support vehicular traffic over poor soil conditions have been used successfully in sand drying beds. In Florida, the Sarasota County Solid-Waste Operations Division used the cellular confinement system, a geosynthetic material made of high-density polyethylene plastic in a range of 2- to 8-in. cell depths. A cross section of the installation consists of a 2-in. layer of sand laid over the sand-filled cellular confinement system, geonet, and then the gravel bed. Cleaning the sand drying bed requires removing only the top 1 to 2 in. of sand that is directly under the sludge, increasing the life of the drying bed.[3]

7.2.2 Wedgewater Filter Beds

The Wedgewater filter bed system was developed in England in 1970. An English group working with the Norwegian pulp and paper industry discovered an efficient sludge dewatering process using a fine wire screen mesh. In 1972 an English group began marketing the system in the U.S. through an agreement with Hendrick

Manufacturing Company of Carbondale, PA. The division has since become an independent company named Gravity Flow Systems, Inc. There are now over 125 operating beds in the U.S. of which nearly 90% are municipal.

The first filter beds used a stainless steel medium called Wedgewire, and a high-density polyurethane medium was developed later. Stainless steel medium has a longer life expectancy than the polyurethane, but is more expensive. They are fabricated in mats 3 ft wide by any length which are laid over concrete floors on structural supports. The high-density polyurethane tiles measure 12 × 12 × 2 in. high and are molded with male and female dovetails for easy installation and removal, as shown in Figure 7-3. The polyurethane media are self-supporting.

The Wedgewater filter system works on the same principle as the sand drying bed, but the Wedgewire or polyurethane creates a capillary action that drains the water more quickly. Its loading capacity of 2 lb dry solids per square foot is about twice that of a sand drying bed, and under favorable conditions the sludge solids content reaches 15 to 20% in about 3 or 4 days on the filter beds, as compared to about 4 weeks on a sand drying bed. Conventional sand drying beds require about 16 times more surface area than Wedgewater beds.

The key to the successful operation of the beds is the preliminary flooding of the bed to just above the medium surface prior to applying the polymer-treated sludge. The drainage rate is then carefully controlled by the operator for the initial period. This provides a hydraulic continuum from the top of the sludge bed to the bottom of the bed; water will then flow more quickly through this saturated profile. As a result, water will drain more quickly than when sludge is applied to a dry or unsaturated surface.

The beds are cleaned with a tractor. The sludge handling system of the Wedgewater drying bed is comparable to that of a belt filter press in terms of polymer injectors and application rates. The costs of sand drying beds and Wedgewater beds are shown in Table 7-1.

7.2.3 Lagoon Dewatering

Lagoon dewatering is a simple, low-cost method of dewatering sludge if the climate is hot and dry, land is inexpensive, and there are no neighbors nearby. Some sludges need to be stabilized to reduce odors before they are lagooned, since lagoons tend to smell more than sand beds. Lagoons are sensitive to weather conditions. In rainy areas, lagooned sludge is slow to thicken. In northern parts of the country, lagoons lose treatment efficiency during colder weather. Sludge lagoons have a relatively large surface area, making odor control difficult. The sludge also may leech into the groundwater, threatening water supplies. These environmental factors, coupled with the odor problems associated with lagoons, have made this an increasingly unpopular choice for sludge dewatering.

Drying lagoons are periodically emptied of sludge and the land refilled. If the sludge layer is 36 cm (15 in.) or less, it will dewater in 3 to 5 months. If the sludge is to be reused for land application, it is typically stockpiled for further drying before application. Sludge will usually not dewater to the point where it can be lifted with a fork except in very hot, arid climates. It is reported that sludge will dry from 5% solids to 40–45% solids in 2 to 3 years using sludge depths of 2 to 4 ft, although definitive data on lagooned sludge are very limited.[2]

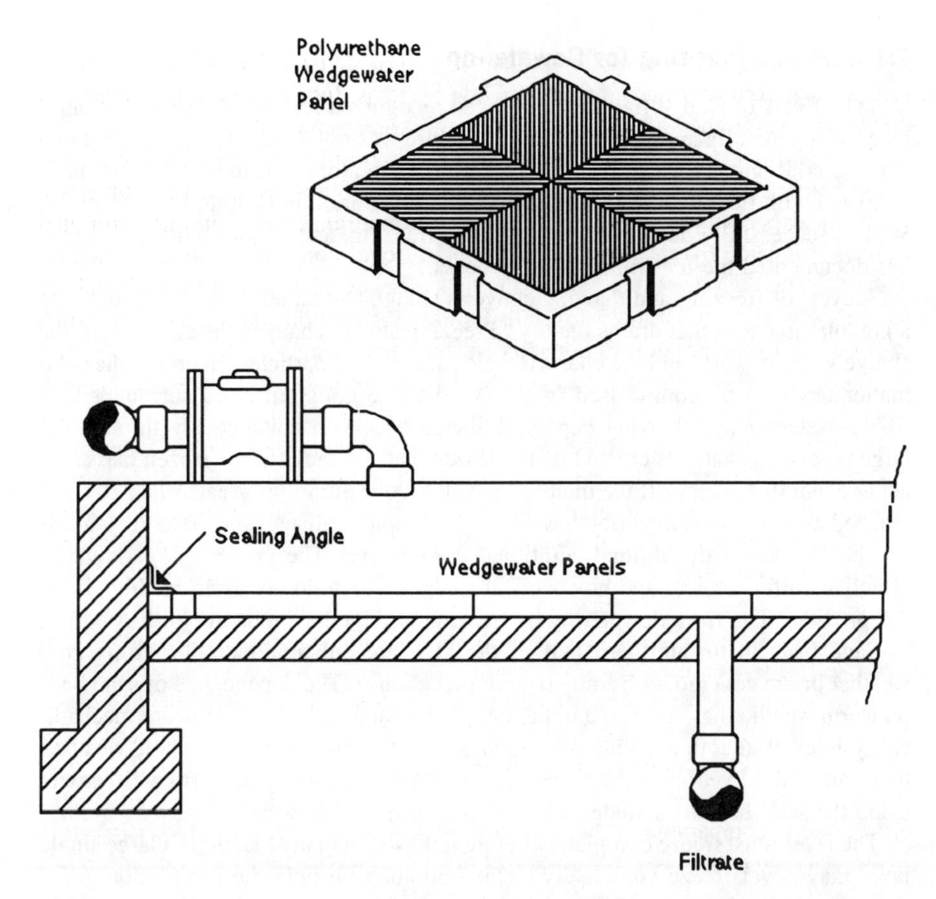

Figure 7-3 Schematic of Wedgewater system in polyurethane and single block detail. (From Wilmut, C. G., Palit, T. T., and Decell, T., *Public Works Mag.* With permission.)

Table 7-1 Cost Comparisons of Natural Dewatering Systems for a 3.8-MGD POTW, Big Spring, Texas

Description	Sand drying beds	Stainless steel Wedgewater beds	Polyurethane Wedgewater beds
Construction costs	$460,000	$538,000	$435,000
Engineering costs	40,000	45,000	38,000
Present worth of O&M	348,000	323,000	338,000
Present worth of alternative	$848,000	$906,000	$811,000

Source: Wilmut, Palit and Decell.[4]
Note: 1990 dollars; based on a facility life of 20 years; 7.625 rate of return.

7.3 INNOVATIVE METHODS OF NATURAL DEWATERING

There are a few innovative methods of natural sludge dewatering. Experimental work on sludge dewatering via freezing has yielded a mathematical model and a pilot facility, but no full-scale projects. Systems that dewater sludge with aquatic plants systems have been installed at over 100 sites in the U.S.

7.3.1 Sludge Freezing for Dewatering

The effects of sludge freezing have been recognized for over 50 years, but until recently a generally applicable design procedure was not available. Recent research at the Cold Regions Research and Engineering Laboratory in Hanover, NH has been validated with field data from a sludge freezing operation in Duluth, MN. Madison, WI has successfully dewatered its sludge by freezing, and similar work in Sweden has documented the feasibility of the process.[5]

A cycle of freezing and thawing converts sludge from a jelly-like consistency to a granular material that drains readily.[6] Freezing sludge changes the structure of the sludge water mixture and the characteristics of the solid particles. In effect, the solid matter tends to be compressed into large discrete conglomerates surrounded by frozen water. When thawing begins, drainage occurs instantaneously through the large pores and channels created by the frozen water. Cracks in the frozen mass also act as conduits to carry off the meltwater. Solids concentrations greater than 20% are reached as soon as the material thaws with very rapid drainage, and 50 to 70% solids can be reached with minimal additional drying time. The process is particularly effective with chemical and biochemical sludges which do not drain readily. Table 7-2 provides data from dewatering by freezing at three locations in North America.

Sand beds in the northern half of the U.S. are either inoperable during cold weather or are covered with some type of greenhouse. The dependence of sand beds on warm weather has sparked a trend away from sand bed systems toward mechanical systems. Rather than taking open sand beds out of operation during cold weather, they can be used year-round by operating them conventionally in warm weather and using the sand beds as a sludge freezing platform in cold weather.[7]

The layer must freeze completely before it thaws. In many locations a large single layer may never freeze completely to the bottom, with only the top portion going through cycles of alternate freezing and thawing. In extremely cold climates such as northern Canada and Alaska, it would be possible to freeze more sludge than can thaw during the short summer season. As a result, the practical sludge depth for arctic and subarctic regions is less than for the northern part of the contiguous U.S.[5]

At most locations in the U.S., sludge application could start by mid-November and end by early March or when the total depth of frozen sludge reaches 100 cm or slightly more. The frozen material should thaw, drain, dry, and be ready for removal by early May so the beds will be ready for polymer-assisted dewatering during the warm months. To ensure successful performance at all times, the design should be based on the warmest winter in the past 20 years and on a layer thickness which will freeze in a reasonable amount of time if freeze-thaw cycles occur during the winter. Recent research has shown that an 8-cm (3-in.) layer of sludge is practical for most locations in moderately cold climates. Figure 7-4 shows the number of inches of sludge that can be successfully dewatered in a winter in the contiguous U.S.[7]

7.3.2 *Phragmites* Reed Beds

Reed bed systems are a specialized application of submerged constructed wetlands for sludge treatment. Conceived at the Max Planck Institute in Germany during the 1960s, the first pilot project with sewage sludge was begun in Usterbach, Bavaria

Table 7-2 Dewatering by Freezing: Sludge Solids Before and After Freezing

Location & sludge type	Percent solids, liquid sludge	Percent solids after freeze/thaw[a]
Ontario, Canada		
Waste activated	0.6	17.0
Anaerobically digested	5.1	26.0
Aerobically digested	2.2	21.0
Hanover, NH		
Anaerobically digested secondary with alum	2–7	25–35
Digested primary sludge	3–8	30–35
Cincinnati, OH		
Waste activated, with alum	0.7	18.0
Water treatment sludges		
With iron salts	7.6	36.0
With alum	3.3	27.0

Source: Adapted from Reed.[7]

[a] As soon as thawing was complete, 1 to 2 weeks additional drying time produced cake of 50% dry solids.

in 1974. To date, there are several hundred systems operating in Canada, Denmark, Sweden, Germany, Great Britain, Austria, New Zealand, and Australia. New York-based Sigmatron Biological Systems pioneered the technology in the U.S. in 1983 and by the end of 1991 had installed over 60 beds in 15 states, with many more scheduled for construction. Aquatic plant systems are considered to be an innovative technology, with demonstrated effectiveness in sludge treatment, but presently without optimized design criteria.[8,9]

The reed bed system takes a traditional sand drying bed for sludge and improves on it. Rather than removing the dewatered sludge from the drying beds after each application, reeds *(Phragmites communis)* are planted in the sand, as shown in Figure 7-5.[10] The sludge can then be added for 7 to 10 years before the beds are emptied. The reed bed is typically constructed of a 40-mil plastic liner covered with a 10-in. bottom layer of 1-in.-diameter stones and a middle layer of 4 to 5 in. of pea gravel. A third layer consisting of 6 in. of sand forms the planting bed for the reeds.[8]

Liquid sludge, aerobically stabilized or anaerobically digested at a solids concentration of 2 to 7% is spread on the surface of the reed bed through a metered system of gravity-fed pipes and troughs. A fully loaded reed bed has about 4 in. of standing liquid on the surface, most of which leaches down through the drainfield in each bed, where pipes return it to the head of the treatment plant. After the liquid has drained a new sludge layer of about 0.5 in. remains, which dries to about 0.25 in. of solids.[8]

The primary function of the reeds is to create a rich microflora in the root zone to feed on the organic material in sludge. *Phragmites* reeds have nodes along the stems for secondary root growth and are resistant to sludge contaminants. The roots extend into the sand and pea gravel, but do not penetrate through the larger stones, so drainage is unimpaired. As new layers of sludge are added to the bed, the *Phragmites* extend their root system through a series of runners or rhizomes. A number of plants will branch off each rhizome and grow vertically through the sludge. The reed plants draw water from the sludge by providing vertical channels through which the water can drain, and water in the sludge is also absorbed and transpired by the reeds. Over time, the sludge will be reduced to about 97% solids.[8]

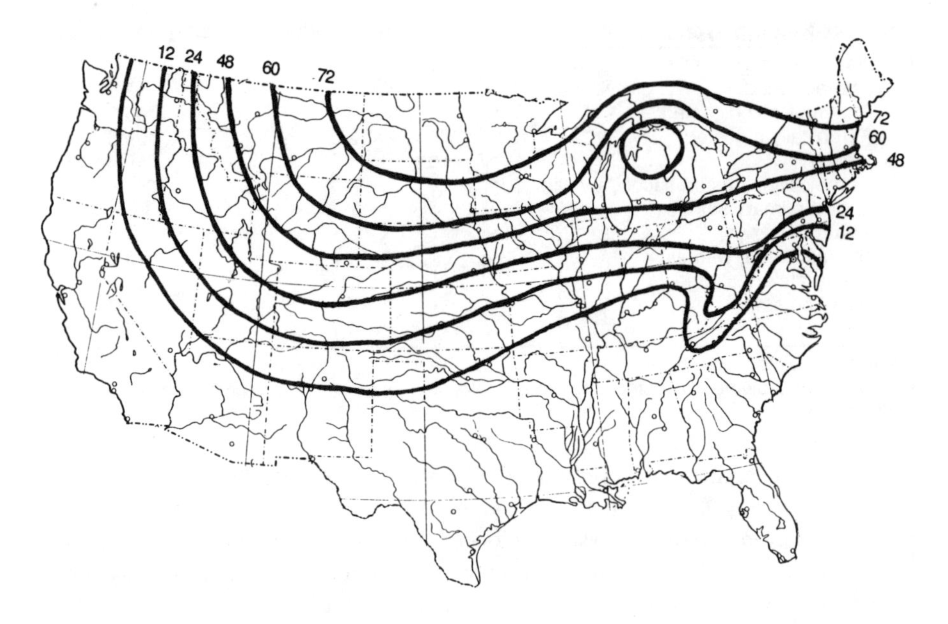

Figure 7-4 Potential depth of frozen sludge if applied in 3-in. layers. (Redrawn from Reed, S. S., *Biocycle,* Jan. 1987. With permission.)

Sludge is applied to the *Phragmites* reed beds year-round, but the reeds are harvested in the fall, leaving their roots systems intact. This plant material may be composted, landfilled, or burned, depending on contaminant concentrations.[9]

After about 7 to 10 years the reed bed is full and no longer can be loaded. It is then taken out of service for about 6 months so the top layer can stabilize before the bed is emptied. The resulting sludge product is a dry, friable material similar to compost or topsoil. It is suitable for land application, provided that the contaminant concentrations are within acceptable limits.[9]

The process is unsuitable for large treatment plants, since piping and drainage are problematic for facilities with capacities of over 5 million gallons a year. The cost of land may be a limiting factor for larger facilities, since a reed bed system requires 1 ft^2 of reed bed for every 20 to 60 gal/day of plant treatment capacity. Finally, the reed bed system is unsuited to southern locations because *Phragmites* need a period of dormancy for winter root growth. Maintenance of the reed beds is quite different from other operation and maintenance tasks at a POTW and includes problems in plant care such as aphid control.[8]

Since none of the beds in this country has been emptied yet, analyses on the end product from a full-scale facility have not been done. Preliminary results from plant analyses and core samples of operating reed beds show some uptake of metals by the stalks and leaves of the reeds and corresponding reductions in metals levels of the sludge product.

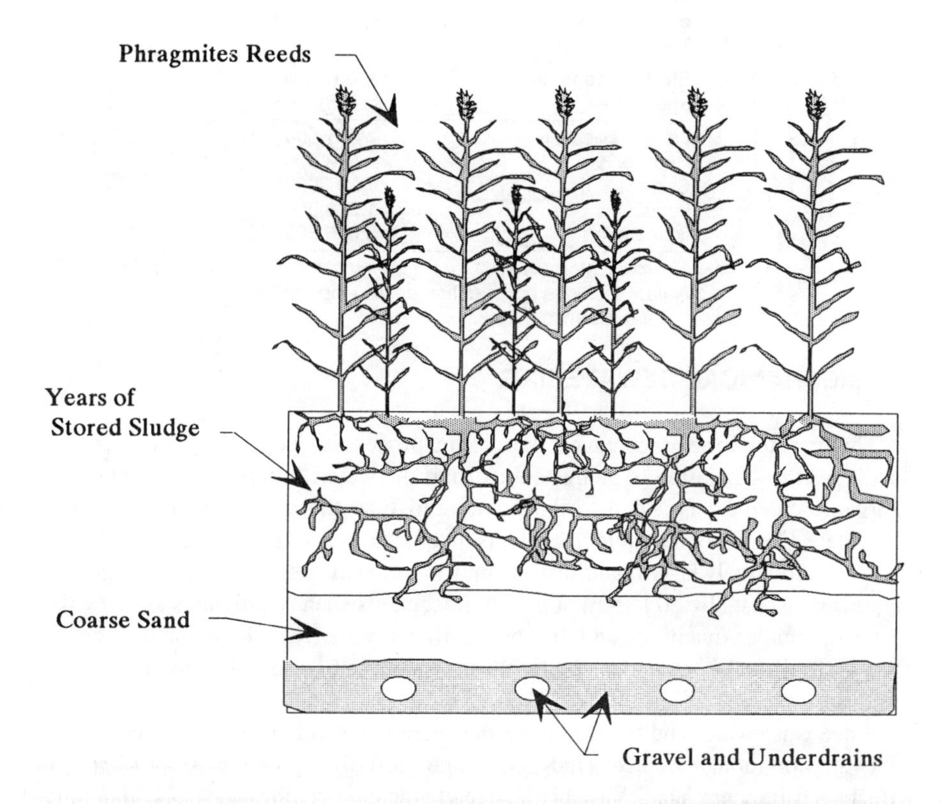

Figure 7-5 Cross section of reed bed. (Courtesy of Susanna Mclwaine.)

Many of the installations thus far have planted reeds in existing dewatering beds, making cost estimates of the system difficult to assess. Table 7-3 provides the cost comparison between a recently completed reed bed system for a 1.4-MGD wastewater treatment plant in Pennsylvania and four other dewatering options. The Borough of Myerstown POTW has 30,000 ft^2 of reed beds that are designed to handle all of the anaerobically digested sludge generated at the wastewater treatment plant. The construction of the reed beds began in the spring of 1990, and they were put into service in October 1990. The installation consisted of six drying beds, each of which measured 50 × 100 ft, for a total of 30,000 ft^2 of bed area. The beds were conventional sludge-drying beds with 1 m of sidewall freeboard. A Hypalon liner was used under the gravel and underdrain layer. The cost of constructing the beds was $623,600, or $20.79/$ft^2$. The cost of installing the reed system in existing sludge drying beds was $67,500, or $2.25/$ft^2$. The cost of installing the reed system in existing sludge-drying beds is currently approximately $3.00/$ft^2$. For calculation purposes it is assumed that when the beds are cleaned the sludge product will be disposed of by landfill. Assuming landfill costs of $60/$yd^3$, the total disposal costs could be approximately $250,000 over the 10-year life of the beds. The disposal costs would be $25,000/year or approximately 4¢ per gallon of sludge applied.[1]

Table 7-3 20-Year Present Worth Analysis of a 1.4-MGD Wastewater Treatment Plant, Myerstown, PA

Sludge management alternative	Present worth ($)
Land application	51,500
Reed beds	508,500
Composting	1,197,000
Landfill	1,578,000
Incineration	3,676,500

Source: McIlwaine.[9]
Note: Includes construction costs + operating expenses, 1989.

7.4 MECHANICAL DEWATERING

The first step in any mechanical dewatering process is sludge conditioning, which facilitates the separation of sludge solids from the water. In this step, an organic or inorganic chemical is added to the liquid sludge so the solids clump together or flocculate. Common inorganic flocculants include lime or ferric chloride, which must be used in relatively large quantities and may increase the total amount of sludge cake requiring disposal by 20 to 50%. Organic flocculants such as polymers are effective in much smaller quantities and do not significantly increase the amount of conditioned sludge cake. Understandably, most new systems use polymers as a sludge conditioner.

Since wastewater and the sludge particles are negatively charged, cationic polymers are commonly used for sludge flocculation. Polymers are more efficient and effective if they are aged. Variable sludge characteristics, including pH, alkalinity, temperature, organic content, solids content, and others, may affect the polymer dosage. After the sludge has been conditioned, either vacuum filters, belt filter presses, or centrifuges may be used.

7.4.1 Vacuum Filter Dewatering

The vacuum filter is among the most common type of mechanical sludge dewatering device. In many places they are being displaced by the more efficient belt filter presses, but at one time there were over 2500 municipal vacuum filters installed around the U.S.

The typical continuous vacuum filter consists of a horizontal drum rotating partially submerged in a reservoir of wet, unfiltered sludge. A filter medium made of various types of material overlays the face of the drum and supports the layer of dewatering sludge. The drum is divided into sectors spanning the length of the drum, each of which is placed under vacuum by automatic valving. As a sector revolves through the reservoir a vacuum is applied, resulting in the formation of a layer of sludge on the filter medium. The vacuum is maintained on the sector as it emerges from the reservoir, resulting in the continuous drainage of moisture from the sludge layer. Drainage continues until just prior to the sector reentering the reservoir. At this point, the sludge cake is automatically removed from the filter medium.

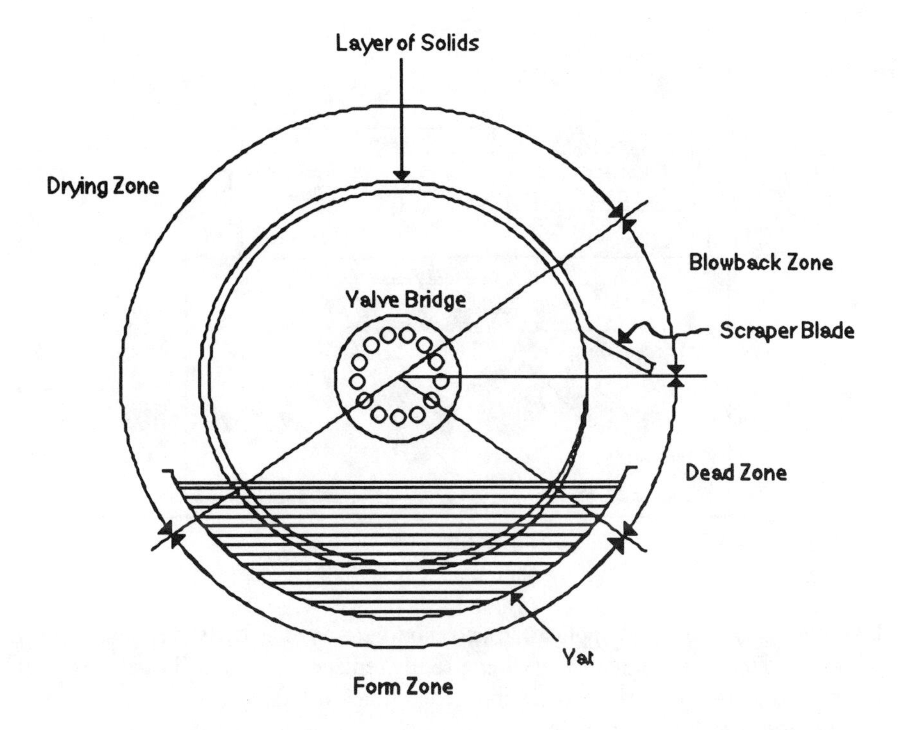

Figure 7-6 Rotary drum filter cycle and discharge. (Redrawn from Eckenfielder, W. W., Jr. and Santhanam, C. J., *Sludge Treatment,* Marcel Dekker, Inc., 1981. With permission.)

The rotary drum vacuum filter uses a synthetic medium which is attached to the drum face. The filter cake, which is formed within the filter vat and dewatered outside the vat, is discharged from the medium by a fixed scraper blade, as shown in Figure 7-6.

As with all continuous rotary vacuum filters, an automatic valving assembly controls vacuum application during the filter cycle. One modification provides for air blowback through the medium just before the scraper to assist cake discharge.

The rotary drum vacuum filter has its greatest application when applied with the concepts of precoat filtration. An expendable precoat medium such as diatomaceous earth is initially formed on the entire drum face to a depth of 5 to 8 cm. A sludge layer is formed on top of the precoat layer in the typical manner of drum filtration and discharged along with a thin layer of precoat by means of a specially designed knife blade. The precoat layer may last up to 24 h of filtration, depending on the application.

One of the major disadvantages of the drum-type filter is the frequency with which the operation must be shut down to wash the medium. Washing is necessary because the sludge tends to progressively blind the medium and impede filtration.[2]

7.4.2 Belt Filter Presses

Belt filter presses were first used to dewater municipal sludges in the early 1970s. The filter press is adapted from the papermaking industry, so the technology has a

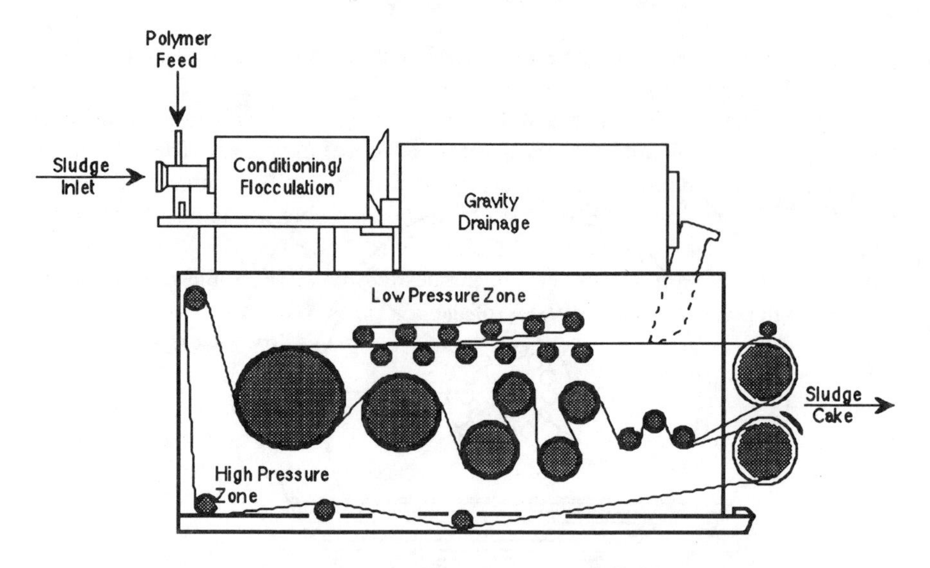

Figure 7-7 Belt filter press. (Courtesy of the City of Omaha's Sludge Reuse Program, 1990.)

long history of industrial application and modification. The early presses produced a drier cake than the vacuum filters they usually replaced, and in the last 20 years belt filter presses have experienced several generations of redesign.

The sludge is processed in three distinct stages: conditioning/flocculation, gravity drainage, and compression shear. A typical layout for the three major process stages of a belt filter press is shown in Figure 7-7.

Conditioning the sludge to agglomerate the suspended solids into a floc is the first step in belt dewatering. Polymer addition and mixing can be carried out in a mixing drum on top of the press, in a separate tank equipped with a mechanical mixer, or by mixing in-line using special pipe sections. The flocculated sludge is then pumped onto a moving porous belt where the free water created during the sludge conditioning drains by gravity, leaving behind a slurry of flocculated sludge solids. The gravity drainage zone can be a rotating screen thickener or, more typically, a long gravity section of belt. The sludge is agitated with plows and contained on the belt as the water drains through it. Gravity drainage after the initial sludge conditioning accounts for 50 to 75% of the water extracted in the mechanical dewatering process and is an essential step before the sludge is squeezed between two belts.[12]

During the compression stage the partially dewatered sludge is squeezed with increasing pressure between two continuous woven fiber belts. The belts are passed over a series of rollers that provide shearing action and increasing pressure as the belts transverse the roller pattern. Water passes through the spaces between the fibers, leaving the solids sandwiched between the belts. At the end of the belt section the dewatered cake is scraped from the belt surface and falls onto a conveyor or other device for further processing or disposal.[11]

Wash water is supplied continuously during dewatering to keep the belt from clogging. The water pressure is usually at 80 psi or more, and the total water flow may be 50 to 100% of the sludge flow rate to the press. It is not necessary to use potable water to rinse the belt; secondary effluent or even recycled filtrate water is of acceptable quality for use as wash water.[12]

Both the roller configuration and the pressure vary by manufacturer. Tension on the belt is supplied by adjusting the position of one of the rollers. The belt tension determines the amount of pressure that will be delivered to the sludge. Belts tend to wander due to variations in cake thickness, so a belt-tracking system is used to sense belt movement. The belt is kept in the center of the rollers by adjusting one end of the tracking roller to move the belt in the desired direction. Some presses have a separate high pressure zone at the end of the belt press where a second set of rollers exerts higher pressures on the sludge to squeeze out additional water.[11]

The processing rate for belt filter presses is usually expressed as a hydraulic loading rate, a convenient measurement since belt presses are usually fed by a positive displacement pump. The loading rate is controlled by varying the pump speed. At low feed-solids concentrations, the capacity of the gravity drainage zone is usually exceeded and belt speed must be slowed. As the feed-solids concentration rises, a point is reached where the solids loading and the cake thickness become the controlling variables, and the loading rate must be decreased so sludge doesn't squirt out the sides of the press. The flocculant and the nature of the sludge solids also affect processing rates. In general, digested sludge is easier to dewater than raw sludge, although it requires more polymer for flocculation.[11]

The parameters that best describe belt filter press performance are the capture efficiency and the cake solids concentration. Capture efficiency usually ranges between 85 and 98%, with a 95% average. Cake solids concentrations vary as a function of the belt press-feed solids concentrations. The American Society of Civil Engineers (ASCE) surveyed over 100 operating belt presses and derived an equation that related cake solids to feed solids concentrations:

$$\text{Percent cake solids} = \frac{\text{percent feed solids}}{0.0915 + 0.0221 \times \text{percent feed solids}}$$

A correlation coefficient of 0.779 was obtained when the equation was fitted to the ASCE data base. Higher correlation coefficients are attainable if the ratio of primary to secondary sludge is more clearly defined.[11]

7.4.3 Centrifuge Dewatering

Centrifugal dewatering uses the rotational force developed by spinning a bowl or basket to separate the sludge solids from the liquids. Disc, basket, and solid bowl centrifuges are all used for sludge dewatering, although the solid bowl centrifuge, or decanter, is the most common. Solid bowl centrifuges are available in either countercurrent or concurrent flow design at either low or high speeds.[13]

The conventional solid bowl centrifuge consists of a solid-walled bowl that rotates at high speed to produce centrifugal forces of 500 to 3000 × g (the acceleration of gravity). Sludge is pumped through a central pipe into a rotating bowl, where the solids are pressed to the inside walls and the lighter liquids pool near the center of the bowl. A scroll, or screw conveyer, inside the machine rotates a little slower than the bowl, and this screw action moves the sludge cake up the inclined beach and out the open end. Centrate is evacuated through holes on either end of the bowl.[13]

A cocurrent centrifuge is shown in Figure 7-8. The centrifuge receives feed sludge and thickens it in the same direction as the flow through the bowl, so the settled solids are not disturbed by the incoming feed. Sludge cake and filtrate pass through the bowl in a smooth, parallel fashion, reducing turbulence. Solids are conveyed over the whole length of the bowl before discharge to provide better compaction and a drier cake and to reduce flocculant requirements.

The centrate flows in the direction opposite the thickened sludge and flows over adjustable weir plates at the feed end of the machine. The horizontal conical conveyor, which is located concentrically within the bowl, rotates at a slightly higher speed than the bowl and conveys thickened sludge to the discharge end of the bowl, where it exits through replaceable nozzles. The differential speed through the bowl and conveyor is controlled by the hydraulic backdrive unit on the conveyor. The hydraulic conveyor drive allows for the adjustment of a preset differential speed that is maintained between the bowl and conveyor. The pool depth is set by adjustable weir plates on the discharge end of the centrifuge.[14]

The countercurrent centrifuge has the feed sludge and the thickened sludge entering and exiting at the same end of the machine. Centrate flows along the axis of the machine and exits along the centrate discharge end. The centrifuge assembly, shown in Figure 7-9, consists of a rotating bowl and conveyor joined through a gear system designed to rotate the two at slightly different speeds.[14] The rotating assembly is covered by a stationary casing for safety and for odor and noise control.

The solid cylindrical bowl is supported between two sets of bearings and includes a conical section at one end. This section forms the dewatering section over which the helical screw pushes the sludge solids to outlet ports and then to a sludge cake discharge hopper. The conveyor has parallel vanes along its axis, interior to the blades at the centrate end of the assembly that clarifies the centrate. The conveyor rotates at a slightly slower speed than the bowl to achieve the differential speed.[2]

The pond depth is controlled by a replaceable disc-shaped weir plate at the centrate discharge end of the assembly that regulates the level of the sludge pool in the bowl. This plate discharges the centrate through the outlet ports by gravity or by a centrate pump attached to the shaft at one end of the bowl. Sludge slurry enters the rotating bowl through a stationary feed pipe and then travels into a baffled abrasion-protected chamber for acceleration before discharge through the feed ports of the rotating conveyor hub into the sludge pool in the rotating bowl. The sludge pool takes the form of a concentric annular ring of liquid sludge on the inner wall of the bowl.[2]

7.4.3.1 Comparing Belt Filter Presses and Centrifuges

Centrifuges do not require a continuous machine wash, and they use much less wash water than belt filter presses. A relatively small amount of water — about 2000 gal per centrifuge per day — is required for the machine wash.

Grit abrasion can cause significant deterioration in centrifuge performance over time. Spinning at high speeds, certain sections of the centrifuge will wear particularly quickly. The feed tube inlet where the sludge is accelerated, the scroll which spins at a different speed than the bowl, and the discharge section where the material is decelerated are especially vulnerable to degradation by grit abrasion.

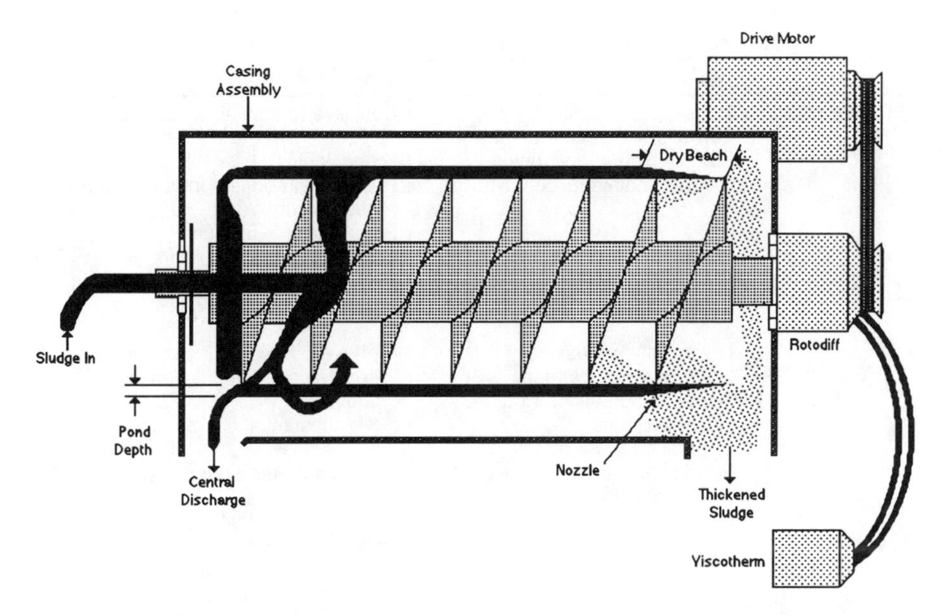

Figure 7-8 Cocurrent centrifuge assembly. (From MacConnell, G. S., Harrison, D. S., Kirby, K. W., Lee, H., and Mousavipour, F., *Water Environ. and Technol.*, Feb. 1991. With permission.)

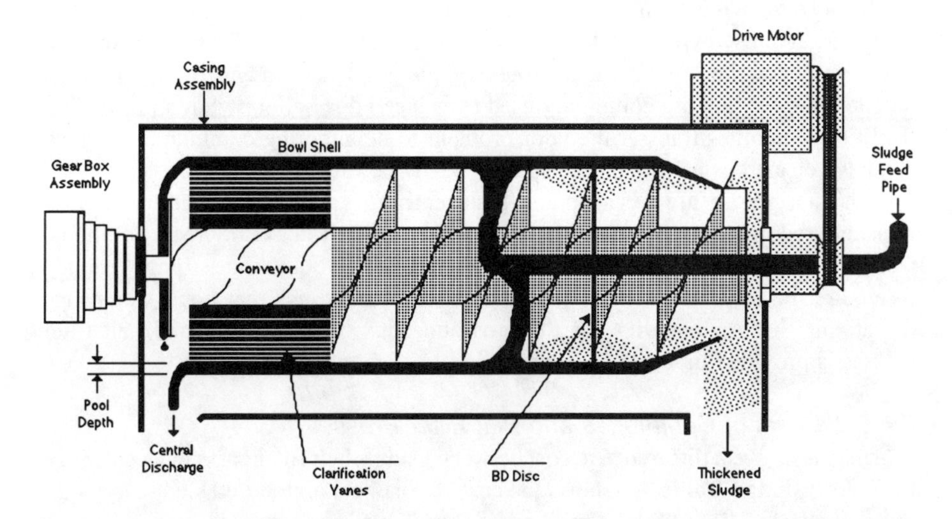

Figure 7-9 Countercurrent centrifuge assembly. (From MacConnell, G. S., Harrison, D. S., Kirby, K. W., Lee, H., and Mousavipour, F., *Water Environ. and Technol.*, Feb. 1991. With permission.)

Rotating centrifuges have substantial vibrations, and the building that houses the centrifuges must be designed to accommodate these stresses. This added structural cost may be significant.

Centrifuges smell less than belt filter presses. When sludge is centrifuged, the odiferous liquids are enclosed and the potential for scenting the atmosphere is

Table 7-4 Advantages and Disadvantages of Dewatering Processes

	Advantages	Disadvantages
Belt filter press	• Lower capital costs • Lower power requirement • Lower noise and vibration • Ease of maintenance	• Sensitive to varying sludge feed characteristics and chemical conditioning • Hydraulically limited in throughput • Higher polymer use • Higher wash water use • Can emit noticeable odors and mist • Requires greater operator attention • Higher maintenance costs
Centrifuge	• Lower polymer use • Lower wash water use • Reduced potential for odor and mist problems • Minimal operator attention • Readily accepts variable sludge feed condition	• High wear if high grit content • Extensive and careful cleaning required • More substantial structural requirements • Requires skilled operator • Noisy • Requires extensive pretesting to achieve optimal machine settings • Change in centrate quality can be easily overlooked because the process is fully contained

Source: Tighe and Bond.[12]

minimized. A small quantity of air is discharged through the dewatered sludge discharge port, which can be conveniently ducted to an odor control device. The sludge filter press exposes a large surface of liquid sludge, offering an excellent opportunity for the transfer of odorous compounds from sludge to air. The high ventilation rates that are commonly used to reduce odor buildup within dewatering facilities sometimes transfer the odor problem to neighboring communities, which can have disastrous effects on the success of the dewatering operation.

Belt filter presses are less adaptable than centrifuges to variable solids concentrations and require more operator attention. However, grit does not create additional wear problems with belt filter presses, and the presses do not vibrate strongly enough to require the structural reinforcement required in buildings that house centrifuge dewatering.[15-17] An overview of the pros and cons of centrifuges and belt filter presses is provided in Table 7-4.

7.4.3.2 Costs of Centrifuges and Belt Filter Presses

There is no great difference in cost between using a belt filter press or a centrifuge to dewater sludge. Table 7-5 shows the costs for using a high liquid sludge throughout centrifuge (240 to 300 gal/min) and a 2-m belt filter press.

7.5 SLUDGE CAKE APPLICATION

The benefits of land-applying sludge cake are equivalent to those of liquid sludge, but the volume is considerably lower and the threat of nitrate leaching is reduced.

Table 7-5 Costs of Using Belt Filter Presses and Centrifuges — 1989

Basis of Design

Sludge type:	Anaerobically digested primary
Typical solids content:	2–4.4%

Design Criteria

	Centrifuge (high throughput)	Belt filter press (2 m)
Cake % solids	27%	29%
Solids capture	95%	97%
Loading rate per		
Unit solids, dry lb/h	3000–5200	1800–2600
Hydraulic, gal/min	240–300	110–180

Approximate Per Unit Costs of Dewatering Sludge

	Centrifuge (high throughput)	Belt filter press (2 m)
Electrical power, $/dry ton	3.7	1.4
Polymer, $/dry ton	$10.00	$16.00
Labor requirements, $/dry ton (at $20/h)	$2.25	$3.75
Maintenance material (initial year, $/dry ton)	$0.65	$1.25

Total Life Cycle Costs for Dewatering Costs over 6 Years at 1020 Dry Tons/Week

	Centrifuge (high throughput)	Belt filter press (2 m)
Capital	$4,996,000	$4,712,000
Labor	843,000	1,414,000
Electricity	1,452,000	522,000
Polymer	3,816,000	6,030,000
Maintenance	238,000	476,000
Salvage value	(–)656,000	(–)500,000
TOTAL	$10,689,000 or $33.60/dry ton	$12,654,000 or $39.75/dry ton

at 400 Dry Tons/Week

	Centrifuge (high throughput)	Belt filter press (2 m)
Capital	$4,996,000	$4,712,000
Labor	544,000	850,000
Electricity	580,000	207,000
Polymer	1,493,000	2,361,000
Maintenance	170,000	306,000
Savage value	(–)656,000	(–)500,000
TOTAL	$7,127,000 or $57.10/dry ton	$7,936,000 or $63.60/dry ton

Source: Tighe and Bond.[12]

Sludge cake is usually applied to farmland with conventional manure spreaders. Different programs use different arrangements for sludge cake distribution: some programs deliver and apply the sludge cake to a site, while others deliver the cake to the farmer for spreading. The permitting of sludge application sites is usually carried out by the sludge reuse program.

Table 7-6 Fertilizer Value of Omaha Sludge

Nutrient	Pounds applied/acre	Price/lb		Value/acre
Nitrogen	500	$0.15		$75.00
Phosphate	200	0.12		24.00
Potash	30	0.25		7.50
Zinc	13	1.00		13.00
			Total	$119.50/acre

Source: Chesnin.[19]

7.5.1 Omaha, Nebraska

Omaha's wastewater treatment plants produce 72,000 tons of sludge each year, 80% of which is directly applied to local farmland as sludge cake. After the sludge is dewatered to 30% solids it has a content of 2.5% nitrogen (0.5% ammonium nitrogen and 2.0% organic nitrogen), 1% phosphorus, 0.15% potassium, and 0.065% zinc. Appealing to the farmers' economic self-interest, the fliers written by the wastewater treatment plant advertise its "Black Gold" as an inexpensive, high-yield fertilizer. The value of the product is presented in Table 7-6.

This does not include all of the benefits, continues the public service advertising. Sludge cake provides organic amendments that improve the physical properties of the soil, a benefit that is difficult to quantify in economic terms. Reducing the density of the soil reduces the fuel requirements of tractors to pull implements through the soil, and sludge-amended crop yields average over 10% higher than yields on the same land using commercial fertilizer. Meanwhile, the farmer will save $30 to $50/acre by making the substitution.

In Omaha's program, local farmers provide the labor and equipment for land application and incorporation of the digested sludge cake, for which the wastewater treatment plant pays them no more than $0.60/yd^3$. Successful bidders typically use their own manure spreaders to apply the sludge cake to their fields.

Application rates are site specific and based on sampling and analysis of the soil, sludge nutrient value, and the crop requirement. Typical sludge application rates are in Table 7-7.

Included in the sludge information packet is the local evidence of the effects of sludge application on crop yields. The information includes the farmer's name, county, acres, crop, and crop yield before and after sludge application, while comments include the number of years that each farmer has used the sludge fertilizer and information on the quality of the soil. This page may be the single most important part of Omaha's sludge reuse program. Interested farmers may contact other sludge users to get an unedited version of the pros and cons of reusing sludge products as fertilizers. In this sparsely populated agricultural state, a farmer's word may be the best advertising that any wastewater treatment plant can have.

Omaha's beneficial reuse program is an undeniable success. The cost of the program, as they reckon it, does not include dewatering the sludge; instead they count only the cost of distributing the product: no more than $0.60/wet ton or $2.00/dry ton.[18] This distribution cost should be added to the cost of dewatering to provide an estimate for the true costs of applying sludge cake to area farmland.

Table 7-7 Application Rates for Omaha, Nebraska Sludge

Crop	Yield goal (bushel/acre)	Sludge applied (wet tons/acre)
Corn	180	15–40
Soybeans	40	0–20
Wheat	50	20–30
Oats	60	20–30
Sorghum	150	40

Source: Anderson.[18]

REFERENCES

1. Girovich, M.J. "Simultaneous Sludge Drying and Pelletizing," *Water Eng. Manage. Mag.* (March 1990).
2. Eckenfelder, W.W., Jr. and C.J. Santhanam. *Sludge Treatment* (New York: Marcel Dekker, Inc., 1981).
3. Banks, J.A., Jr. and W.K. Lederman. "Innovative Sludge Drying Bed Design," *Public Works Mag.,* (September 1990).
4. Wilmut, C.G., T.T. Palit, and T. Decell. "Sludge Made Manageable with Wedgewater Filter Beds," *Public Works Mag.,* (October 1990).
5. Reed, S., J. Bouzoun, and W. Medding. "A Rational Method for Sludge Dewatering via Freezing," *J. Water Pollut. Control Fed.* 58(9):911–916 (1986).
6. "Developments in Dewatering," *Biocycle* (November/December 1986).
7. Reed, S.C. "Sludge Freezing for Dewatering," *Biocycle* (January 1987).
8. Riggle, D. "Reed Bed System for Sludge," *Biocycle* (December 1991).
9. McIlwaine, S. "Introduction to Constructed Wetland and Aquatic Plant System Wastewater Treatment for Small Northern Communities," Prepared for Vermont's Rural Housing, Inc. (August 15, 1990).
10. Sassaman, M.D. and T.R. Kaufman. "Sludge Dewatering and Disposal Utilizing the Reed System," Presented at the Water Environment Federation 65th Annual Conference and Exposition, New Orleans, LA, September 20–24, 1992.
11. Task Committee on Belt Filter Presses. "Belt Filter Press Dewatering of Wastewater Sludge," *J. Environ. Eng.* 114(5):991–1005 (1988).
12. Tighe and Bond, Inc. "Interim Sludge Processing and Disposal Project, Final Facilities Plan, Environmental Impact Report," Vol. 2, prepared for the Massachusetts Water Resources Authority (April 1989).
13. "Sludge Thickening Manual of Practice No. FD-1," Water Pollution Control Federation, Washington, D.C. (1980).
14. MacConnell, G.S., D.S. Harrison, K.W. Kirby, H. Lee, and F. Mousavipour. "Centrifuge vs. Dissolved Air Flotation," *Water Environ. & Technol.* (February 1991).
15. "Developments in Sludge Dewatering," *Biocycle* (March 1986).
16. Coker, C.S., R.L. Walden, T.G. Shea, and M.J. Brinker. "Dewatering Municipal Wastewater Sludge for Incineration," *Water Environ. & Technol.* (March 1991).
17. Donovan, K. "City Cuts Sludge Dewatering Costs," *Public Works Mag.,* (July 1989).
18. Anderson, G. "General Program Information," City of Omaha's Sludge Reuse Program, Quality Control Division (1990).
19. Chesnin, L. "Omaha's Gold," *Soil Science News,* Cooperative Extension Service, Institute of Agriculture and Natural Resources, University of Nebraska, Volume VI, No. 3 (March 30, 1984).

8 LIME-STABILIZED SLUDGE

8.1 INTRODUCTION

Lime stabilization has been used for centuries to reduce the odor generation and pathogen levels of biological matter. It is a simple and effective method of sludge processing that requires little specialized equipment. When lime is mixed with dewatered sludge, microorganisms involved in the decomposition of the sludge and pathogens are inhibited or destroyed. With a proper level of lime stabilization, biological activity is virtually halted and there is little or no decomposition. Since odors are produced from the biological emissions of gases, lime stabilized sludge doesn't smell. The sludge product produced by this process is typically a granular, soil-like substance with a faint odor of ammonia.[1]

The principal advantages of lime stabilization are that the process is relatively simple to operate and the capital requirements are low. The disadvantage is that it increases the weight and volume of the sludge product to be distributed, raising the cost of sludge transport. The low fertilizer value per volume and high weight of lime-stabilized sludge restrict its application to agricultural sites within a short distance of the POTW. However, lime is routinely land applied by farmers to raise the soil pH and "sweeten" their fields. Lime-stabilized sludge is not only odor free and biologically inactive; the lime content is a boon to farmers and makes the product more attractive to potential land applicators.[2]

For lime-stabilized sludge to meet class A standards, the pH of the sewage sludge must be raised above 12 and remain there for at least 72 h. During this time, the temperature of the sludge must be above 52°C for 12 h or longer. At the end of the 72-h period during which the pH of the sludge is above 12, the sludge is air dried to greater than 50% solids. For a class A sludge product, in addition to lime treatment either the density of fecal coliforms in the sludge must be less than 1000 most probable number (MPN) per gram of total dry solids or the density of *Salmonella* sp. bacteria in the sludge must be less than 3 MPN/4 g of total dry solids at the time the sludge is used or sold. Class A standards are usually met through lime pasteurization or chemical fixation.[3]

For lime-stabilized sludges to meet class B standards, the pH must be raised to 12 or higher by alkali addition and shall remain at a pH of 12.0 or higher for 2 h without the addition of more alkali and then at 11.5 or higher for an additional 22 h. The geometric mean density of fecal coliforms shall be less than 2 million MPN or colony-forming units (cfu) per gram of total solids (dry weight basis).[3]

A sludge that has been lime stabilized to a pH of 12 or above achieves significant pathogen reduction. The high pH liberates ammonia into the sludge air space. Ammonia is an excellent disinfectant and contributes to the overall microbial control and suppression of microbially produced odors. Table 8-1 gives the pathogen reduction levels of lime-stabilized sludge.

As shown in Table 8-2, the lime stabilization of most sludges results in significantly lower levels of pathogens than anaerobic or aerobic digestion. Composted sludge shows similar reductions in fecal coliforms and fecal streptococci.

8.2 LIME STABILIZATION CHEMISTRY

Lime is commercially available in two forms: quicklime and hydrated lime. Direct addition of dry quicklime without slaking (that is, using water in a ratio of 3:1 or more to produce a wet hydrate) is a common practice. The advantages of this approach are that slaking equipment is eliminated and that the heat generated by the reaction of the quicklime with the entrained water is used to improve pathogen reduction. The particle size of the lime affects the rate of hydration and, therefore, the rate of pH change which is used in the design of lime-sludge mixing equipment. A pulverized lime will improve the hydration rate, but may adversely affect the storage and feeding equipment.[4]

The sludge pH is raised by adding calcium hydroxide ($Ca(OH)_2$). Quicklime (CaO) must first be converted to calcium hydroxide by hydration to raise the sludge pH. The chemical reaction that occurs when quicklime is mixed with water is expressed as the hydration reaction shown in Equation 8-1:

$$CaO + H_2O \Rightarrow Ca(OH)_2 + \text{heat} \quad (8\text{-}1)$$

Dolomitic lime, when hydrated under atmospheric conditions, gives Equation 8-2:

$$CaO \cdot MgO + H_2O \Rightarrow Ca(OH)_2 \cdot MgO + \text{heat} \quad (8\text{-}2)$$

Each pound of 100% quicklime produces 491 Btu of heat and 1.32 lb of calcium oxide and extracts 0.32 lb of free water from the sludge. Each pound of 100% dolomitic lime under atmospheric conditions combines with the water to produce 380 Btu of heat and 1.2 lb of calcium oxide and extract 0.18 lb of free water from the sludge.[4]

Precise reaction equations cannot be written with a heterogeneous mixture such as sludge, but a generalization can serve as the basis for design criteria. A 1-MGD POTW will produce about 1 ton of dry solids per day. These solids will be dewatered to a sludge cake of about 20% solids. The lime required to raise the pH of this cake to 12.0 or higher can be determined accurately by testing, although experience shows that the lime demand may vary from 0.25 to 2.0 lb per pound of dry solids. A 1-MGD POTW will usually require somewhere from 0.5 to 1.5 ton of CaO per day.

Table 8-1 Pathogen Reduction in Lime-Stabilized Sludge

Lime dose (% lime)	Fecal coliforms (MPN/100 ml)	Pathogen reduction (%)
0.0	23,000,000	0.00
6.0	93,000	99.60
8.0	93,000	99.96
12.5	4,300	99.98
16.0	2,300	99.99
25.0	430	99.99
50.0	430	99.99

Source: Ritter.[12]

Table 8-2 Comparing Log Reductions of Pathogens by Different Sludge Processes

	Log reduction	
Process	**Fecal coliforms**	**Fecal streptococci**
Anaerobic digestion[a]		
Mean	1.84	1.48
Range	1.44–2.33	1.10–1.94
Composting[b]	≥4	2.9
Aerobic digestion[c]		
20°C	1	1
30 and 40°C	2	2
Lime stabilization		
Standard process		
Raw primary[d]	5.1	3.4
Raw primary[e]	3.8	3.3
Raw primary[h]	5.1	2.4
Vacuum filtered[h]	5.8	3.0
Fe^{3+} conditioned primary[f]	4.0	2.0
Mixed primary[g] and trickling filter humus, 2% solids	7.2	3.4
Mixed primary[g] and trickling filter humus, 4% solids	2.6	1.8
Waste activated[d]	3.2	3.2
Anaerobically digested[d]	2.6	1.5
Cake and dry lime[h]	5.2	2.7

Source: Westphal and Christensen.[13]

[a] Full-scale studies, 14- to 15-day retention time, 35°C.
[b] Full-scale studies, 9- to 28-day retention time, 15 to 70°C.
[c] Laboratory study, 35-day retention time.
[d] Full-scale study, pH raised to 12.4, tested shortly after treatment.
[e] Laboratory study, pH raised to 12.5, tested shortly after treatment.
[f] Laboratory study, pH raised to 12.5, tested 0.5 h after treatment.
[g] Laboratory study, pH raised to 12.4, tested 1 h after treatment.
[h] Laboratory study, pH ≥ 12 at 2 h, tested 2 h after treatment; values shown are averages of 6 tests.

8.3 LIME STABILIZATION PROCESSES

Three lime-sludge mixing systems have been developed. Liquid sludge can be mixed with lime and land applied by tank trucks, it can be mechanically dewatered after the addition of lime, or lime can be mixed with sludge cake that has been dewatered to about 20% solids. Addition of lime to dewatered cake is known as post-dewatered lime stabilization and is the most feasible method of lime stabilization.

8.3.1 Sludge Feed Equipment for Lime Stabilization[5]

Screw conveyors are generally the best type of conveyor to handle sludge.[5] Depending on the sludge characteristics, the actual capacity of the screw conveyor may be as little as 15% of the Conveyor Equipment Manufacturers Association (CEMA) published data. In addition, the revolutions per minute and the tip speed of the conveyor play an important part in determining the capacity and conveyability of the sludge. A manufacturer experienced in handling municipal sludge can properly evaluate the sludge characteristics and recommend a size and operating speed to ensure capacity and prevent a thixotropic condition from occurring.

Screw conveyors are ideal for adding supplemental heat to preheat the sludge prior to lime stabilization. By using supplemental heat, an average lime reduction of 18% has been noted. The heat may be supplied by electricity, hot water, or steam. The screw conveyor heat source should be insulated to optimize efficiency and provide a safety factor.

8.3.2 Lime Storage and Feed Equipment[5]

Lime is usually purchased in bulk, and the lime storage equipment[5] should be sized to hold at least $1^1/_2$ full truck loads and a maximum of a 1-month supply of lime. The storage equipment should protect the contents of the silo to prevent moisture from entering and degrading the lime. Figure 8-1 illustrates a recommended lime storage and feed system.

Lime delivery trucks are equipped with a blower and 4-in. quick-connect fittings for attachment to the silo. The lime silo should have an exhaust fan to remove the air being blown into the silo by the truck blower and a filter to remove dust from the air. There are two suitable dust collectors for this purpose: a pulse cleaning type that cleans the filter cloth while the exhaust fan is running, and a mechanical shaker type which requires that the fan be turned off before cleaning the cloth. A mechanical shaker type filter is better suited to this application, since the filter is only in operation when the silo is being filled. The dust collector exhaust fan should provide roughly 1000 ft^3/min capacity and about 240 to 250 ft^2 of cloth area.

The fill line should have a 4-in. male quick-connect coupling for connection to the delivery truck, and a 4-in. schedule 40 pipe should be used with 48-in.-radius elbows to the top of the silo. The top of the silo should be fitted with a center target box to dissipate the velocity of the lime and to provide a cleanout cover.

Upon delivery, the initial aerated density of the lime is lower than the published volumes. The volume of the silo should be based on the aerated bulk density and not the settled density, as shown in Table 8-3. Typically, 12-ft-diameter silos provide the most economical volume. The minimum dimensions recommended for a silo are a 12-ft diameter, a 16-ft straight sidewall height, and a 60° conical bottom.

The lime feed equipment train starts with a vibrating bin activator which is bolted to the bottom of the silo. The bin activator diameter is generally one half the diameter of the bulk storage silo, so a 12-ft-diameter silo requires a 5- or 6-ft-diameter

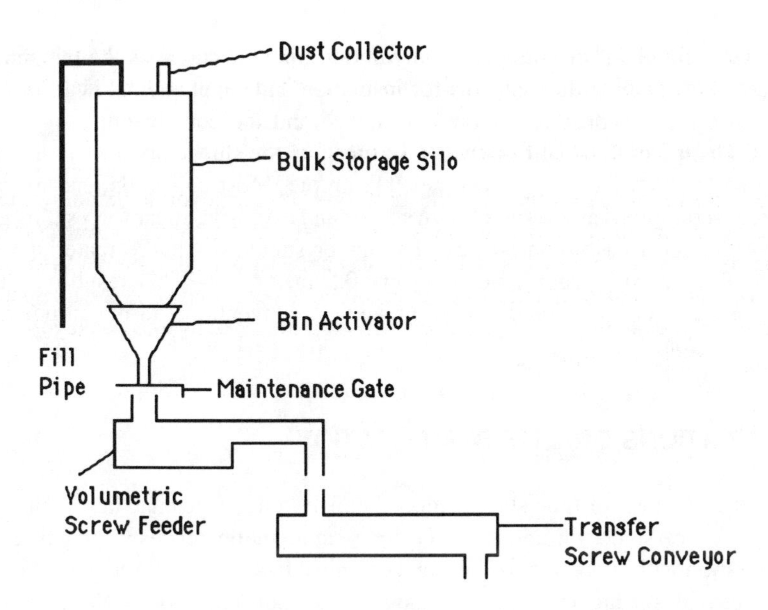

Figure 8-1 Lime storage and feed equipment. (Courtesy of the RDP Company.)

Table 8-3 Bulk Density of Lime: Calculating Silo Size

	Bulk density	
	Settled	**Aerated**
Pebble lime	55–60	45
Pulverized lime	70–75	60
Hydrated lime	35	25

Source: Christy.[5]

activator to provide good operation. The bin activator outlet should have a maintenance gate to block off the lime from the downstream equipment. Knife gate valves are not appropriate for this application since they vibrate apart after a short period.

The bin activator ensures a constant feed to the volumetric feeder. Either a rotary valve feeder or a volumetric screw feeder can be used, although the screw feeders typically require less maintenance, experience less abrasion, and are generally more reliable than the rotary valve feeders.

8.3.3 Lime/Sludge Mixing Equipment[5]

The mixer[5] must blend the sludge and lime thoroughly to ensure that the pH is 12 or greater at the discharge. The mixer should be heated and insulated to benefit from the savings in lime from supplemental heat. The insulation will retain both the supplemental heat and the heat generated by the exothermic reaction of the lime reacting with the entrained water in the sludge.

The mixer should provide a minimum retention time of 60 to 90 s. The lime generally takes 60 s to hydrate, since the mixer ensures that the lime is uniformly distributed throughout the sludge. The hydration time decreases as the temperature increases, providing further impetus for insulation and supplemental heat.

The mixer is responsible for the appearance and the conveyability of the end product. The mixer must be fast enough to break up the sludge and mix in the lime, but not so fast that the sludge becomes thixotropic. Most pug mills tend to create thixotropic conditions as a result of high speeds and excessive shear forces; this gives the end product a pasty consistency. The mixer should discharge material which contains 95% small particles of diameters 0.5 in. or less. This resulting sludge product is easily conveyed and can be spread with conventional farm equipment for land application.

8.4 VARIATIONS ON LIME STABILIZATION

The effectiveness of lime stabilization chemistry on sludge cake has inspired the development of some variations of the lime stabilization process, including lime pasteurization with supplemental heat and chemical fixation. The lime pasteurization process provides a higher level of pathogen reduction than lime stabilization and typically will result in a sludge product that meets the class A requirements. Lime pasteurization with supplemental heat reduces the amount of lime necessary for pasteurization by additional heat input. Chemical fixation uses lime kiln dust (LKD) or cement kiln dust (CKD) rather than lime and produces a sludge product that is a friable, structurally sound earth substitute.

8.4.1 Lime Pasteurization

In lime pasteurization, lime is added to the sludge and a pH of 12 at 70°C is maintained for 30 min. The high temperature can be achieved by the addition of lime alone or by using less lime and adding supplemental heat. The use of supplemental heat can be more cost effective than using additional lime.

The lime pasteurization process is shown in Figure 8-2. After thoroughly mixing the lime and sludge, the mixture is transferred to a pasteurization vessel which is designed to maintain a sludge temperature of 70°C, provide a minimum detention time of 30 min, and provide documentation showing temperature vs. time. One or more vessels may be used, depending on the layout and design capacities. Pasteurization vessels should be insulated to retain the heat of the sludge. A small amount of heat tracing on the vessel walls is used to overcome the heat loss and reduce operating costs.

The formula shown in Table 8-4 can be used to predict the quantity of lime required to pasteurize sludge. The example shown is based on 1 ton of sludge cake at 18% dry solids.

In RDP Company's patented "Envessel Pasteurization" system, the dewatered sludge is transferred by a double screw conveyor which feeds the sludge into a second double screw conveyor where it is mixed with lime. Both the sludge feed conveyor and the sludge/lime mixture are covered with thermostatically controlled heat blankets. After mixing, lime/sludge combination is fed into a two-sided pasteurization vessel. As one side cures for 30 min at a pH of 12 and 70°C, the other side

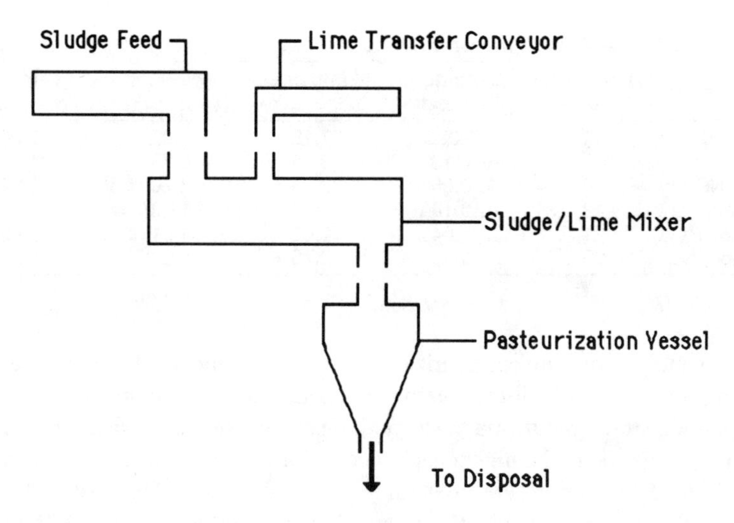

Figure 8-2 Lime pasteurization process. (Courtesy of the RDP Company.)

Table 8-4 Worksheet for Lime Pasteurization: 2000 lb Sludge Cake at 18% Dry Solids

1. Calculated heat required to elevate temperature from 60°F to 160°F:

pounds	×	specific heat	×	temperature rise	=	Btu
360	×	0.30	×	90°F	=	9,720
1,640	×	1.0	×	90°F	=	147,600

2. Calculate lime required based on 490 Btu/lb of CaO:
   157,320/491 = 320 lb CaO or 345 lb lime @ 93% CaO

3. Approximate cost:
   345 lb of lime at $75/ton = $12.90/ton cake

4. Reduce lime addition to 30% on a dry solids basis:
   360 lb dry solids × 0.30 = 180 lb lime

5. Heat available from lime:
   108 lb × 93% CaO × 491 Btu/lb = 49,306 Btu

6. Supplemental heat required (from Step 1):

	157,320 Btu
	– 49,316 Btu
	108,003 Btu

7. Cost per ton of cake:

108 lb lime @ $75/ton	$4.05
108,003 Btu @ 0.075/kWh	+$2.43
Total	$6.48

Source: Christy.[5]

completes the 30-min pasteurization period, empties, and refills. Both sides of the pasteurization vessel are fitted with double screw bottoms and a thermostatically controlled heat blanket.[7]

Table 8-5 Pilot Lime Pasteurization Results

Location	Sludge feed (% solids)	Lime dose (lb/lb cake)	Lime cost ($/ton cake)	Heat cost ($/ton cake)	Total cost ($/dry ton cake)
Topeka, KS	24	0.095	7.12	1.06	34.08
Aiken, SC	22	0.13	9.75	0.75	47.73
Durham, NC	19	0.05	3.75	2.76	34.26
Clearwater, FL	12	0.10	7.86	1.49	77.92
Cleveland, TN	21	0.04	2.60	1.61	20.05
Wilmington, NC	25	0.05	3.75	2.38	24.52

Source: Christy.[5]

The advantages of lime pasteurization with supplemental heat are the quick processing time and the reduced rate of lime addition. The shorter processing time increases sludge throughput, and the lower lime content reduces transportation costs. The disadvantages lie in the higher capital costs for the pasteurization vessel and the increased energy inputs during processing. Table 8-5 shows the cost of lime and supplemental heat for pilot facilities in six cities around the country. Although the costs for lime pasteurization appear to be quite low in this analysis, a more complete accounting for this process is presented in Table 8-6.

8.4.2 Chemical Fixation of Sludge

Chemical fixation emerged in the 1970s as a method of managing hazardous wastes. It usually is used to improve the handling and physical characteristics of the waste, decrease the pollution-transfer surface area, or limit the solubility of hazardous substances contained in the waste. When the chemical fixation process is applied to sludge, it is a modified lime stabilization process: quicklime and an additive are processed with the sludge to produce a friable, soil-like material that is odorless and easily handled. Additives include portland cement, sodium silicates, CKD, and LKD.[8]

The addition of CKD and LKD to the quicklime used in lime stabilization are variations of the lime stabilization process that are patented by N-Viro Systems of Toledo, OH and Chemfix Inc., respectively. CKD and LKD are by-products of the cement or lime manufacturing process. The product is essentially fly ash and some amount of calcium oxide, and the metal of particular concern is lead, which precipitates out with kiln dust. Kiln dusts contain lower concentrations of calcium oxide than high-calcium lime. Instead of the 1:5 ratio of lime to sludge solids used in lime stabilizing dewatered cake, N-Viro and Chemfix use a ratio of roughly 1:1 or 2:1 of CKD or LKD and quicklime:sludge solids, bulking the sludge up considerably. This reduces the nutrient value of the end product.[9]

Table 8-7 shows the nutrients and metals concentrations of lime stabilization and chemical fixation of a Durham, NC sludge. With lime stabilization, the metals concentration of the sludge product is always lower than the metals level in the sludge feed. With the addition of CKD, three out of ten metals show higher concentrations in the CKD-stabilized sludge product than in the sludge feed.

The chemical fixation process starts by mixing the dewatered sludge with CKD or LKD and quicklime in a pug-mill mixer. The material is held in a concrete bunker for a curing period of about 12 h. The mixture is then windrowed, where it dries to a solids content of 50% in 3 to 7 days.

Table 8-6 Costs of Lime Pasteurization and Chemical Fixation in Burlington, VT

	Lime Pasteurization		Chemical Fixation	
	Annual costs	$/dry ton processed	Annual costs	$/dry ton processed
Operating expenses				
Direct labor and benefits	$60,000	$22.12	$60,000	$22.12
Electricity	94,000	34.66	113,000	41.67
Fuel	4,000	1.48	18,000	6.64
Chemicals	91,000	33.56	559,000	206.12
Maintenance and repairs	49,000	18.07	27,000	9.96
Equipment replacement	15,000	5.53	8,000	2.95
Royalty	0	0	43,000	15.86
Miscellaneous	31,000	11.43	83,000	30.61
Total operating expenses	344,000	126.85	911,000	335.91
Annualized construction costs (60%)	206,400	76.11	390,800	114.10
Net costs	$550,400	$202.96/dry ton	$1,301,800	$450.01/dry ton

Note: Based on 2712 dry tons processed, fiscal year 1993, state subsidy = 40% of construction.

Source: Dubois and King.[7]

Table 8-7 Metals and Nutrient Concentrations for Lime-Stabilized and Chemically Fixed Sludge

Metal/nutrient	Sludge feed (mg/kg)	Lime stabilized (mg/kg)	CKD[a] stabilized (mg/kg)
Arsenic	7.5	4.5	8.5
Cadmium	3.5	2	8.5
Chromium	53.5	27.5	48
Copper	665	262.5	215
Lead	58	44	570
Mercury	3.7	1.6	1.45
Molybdenum	88	56	78
Nickel	24.5	14	18.5
Selenium	25.5	7	8
Zinc	1900	772.5	650
Alkalinity, $CaCO_3$	6,150	262,500	42,000
Phosphorus	25,500	10,825	7,850
Potassium	2,350	1,250	15,000
Kjeldahl nitrogen	8,800	2,150	825
Ammonia nitrogen	5,700	1,350	245
Nitrate	5	11	2
Plant available nitrogen	6,326	1,521	364
Total solids (percent)	19.4%	41.1%	50.9%

[a] CKD = cement kiln dust.

Source: Christy.[5]

When dewatered sludge is added to the anhydrous cement powders, a colloidal calcium-silicate-hydrate gel is produced. The gel hardens to form thin, interlaced, densely packed silica fibrils that incorporate the aggregates and waste into a crumbly, soil-like material called Chemsoil (by Chemfix) or N-Viro Soil. The process is particularly effective for toxic metals because the pH of the cement mixture is high enough that most multivalent cations are converted into insoluble hydroxides or carbonates or are directly incorporated into the cement minerals that form. Soluble silicates assist in the binding of materials in the cement solidification process, serving to precipitate most interfering ions and speed setting. Research suggests that soluble silicates can stabilize sludge containing high concentrations of heavy metals under very acidic conditions.[8]

The sludge product produced by these processes could be land applied theoretically, but both companies have had limited success in securing land application sites, since farmers generally don't want to put CKD or LKD on their fields. Disposal of the material can be problematic; it is usually used as a daily cover for landfills.

When chemically fixed sludge product is used as landfill cover, it is typically blended with existing on-site soils to improve its workability and is then placed as a daily cover on the working face of the active landfill area. The mix ratio of natural soil to artificial soil varied from 5:1 to 2:1 in a project carried out by the McAlpine Creek Wastewater Management Facility of Pineville, NC, which attributed the range to variations in alkaline material type and dosages which improved the workability of the product for daily cover applications.[10]

The most attractive attribute of chemical fixation is that either company can set up a mobile sludge processing facility at a POTW in a matter of weeks. This quick response time offers an immediate solution to a pressing sludge management problem. One set of cost estimates for lime pasteurization and chemical fixation is provided in Table 8-6. These costs do not include the sludge dewatering costs.

8.5 LIME-STABILIZED SLUDGE APPLICATION

The effects of land-applying lime-stabilized sludge are similar to those of dewatered sludge cake except for the inclusion of lime, an additive that many farmers find attractive. The lower levels of pathogens and biological activity make the sludge product easier to manage, although the increased weight and volume add to transportation costs. Like dewatered sludge, lime-stabilized sludge is usually applied with conventional manure spreaders.

8.5.1 Oklahoma City, Oklahoma

Oklahoma City began land-applying liquid sludge in 1982. By 1987 they put their sludge program up for competitive bid in an effort to reduce costs. The recipient of the contract, Professional Services Group, Inc., is a private operator that uses post-dewatered lime stabilization to process the sludge, which is locally land applied.

The sludge is dewatered by belt presses and conveyed to a ribbon blender/pug mill where lime is added and mixed with the sludge so it reaches a pH of 12.0 or greater for at least 12 h. After curing, the solids content of the sludge product is about 30%, and it is trucked to the application site and land applied.

The largest savings have been in sludge trucking. Oklahoma City had been land-applying liquid sludge with 6500-gal tanker trucks, which took 60 to 65 round truck trips per day 7 days a week. With the higher solids content of the lime-stabilized sludge, truck trips have decreased to 18 to 20 per day, 5 days a week.

The sludge cake is transported to farm sites within a 10-mile radius of the treatment plant, and over 12,000 acres are permitted under the program. Front-end loaders fill the dump trucks with sludge product for transport to the application sites, where dry sludge applicators are used to apply the material at an even rate. With 24 h of application the limed sludge is disked 6 to 8 in. into the soil.

The application rate is determined by the plant-available nitrogen (PAN), consumptive use of the proposed crops, the existing soil nitrogen, and an annual cadmium application limit of 0.446 lb/acre. Soil samples from the application sites are annually tested for PAN, cation exchange capacity, pH, cadmium, and any other constituents that were considered to be important from the initial analyses.

Farmers who participate in the Oklahoma City sludge program do well. The tillage value of the program is about $10/acre, while an equivalent amount of commercial fertilizer would cost about $76/acre for available nitrogen, $6/acre in phosphate, and $10/acre for potash; hence, the city gives area farmers benefits worth $102/acre for participating in the program. The organic material in the sludge product increases the ability of the soil to retain moisture, and a number of farmers have reported a yield increase of up to 20 bushels/acre.[11]

REFERENCES

1. Burnham, J.C. "Reduction of Odors in Kiln Dust Treated Wastewater Sludge Cake," *RCRA Rev.* 3(3):4–15 (1991).
2. Roediger, H. "Using Quicklime: Hygienization and Solidification of Dewatered Sludge," *Operations Forum* (April 1987).
3. "40 CFR Part 503 Sludge Regulations," U.S. Environmental Protection Agency (1993).
4. Christy, R.W., Sr. "Sludge Disposal Using Lime," *Water Environ. and Technol.* (April 1990).
5. Christy, P.G. "Process Equipment Considerations for Lime Stabilization Systems Producing PSRP and PFRP Quality Sludge," RDP Company, prepared for the Water Pollution Control Federation Conference "The Status of Municipal Sludge Management for the 1990s," New Orleans, LA (December 1990).
6. Christy, P.G. "Lime Pasteurization, An Extended Evaluation," RDP Company (1992).
7. "Sludge Management Plan Draft Report," Prepared by Dubois and King for the Chittenden Regional Management District (1991).
8. Mulbarger, M.C., R.P. Trubiano, T.L. Papes, and G.L. Gallinaro. "Past Practices and New Approaches in Scum Management," *Water Environ. Technol.* (November 1990).
9. Burnham, J.C., N. Hatfield, G.F. Bennett, and T.J. Logan. "Use of Kiln Dust with Quicklime for Effective Municipal Sludge Pasteurization and Stabilization with the N-Viro Soil Process," Standard Technical Publication 1135, American Society for Testing and Materials (1992).
10. Mendenhall, T., D.H. Moreau, N.V. Colson, and L.A. Stone. "Artificial Soil — It's Use As Landfill Cover Material," unpublished.
11. Couch, J. and E. Tacha. "Cutting Costs for Sludge Management," *Biocycle* (May 1989).
12. Ritter, D. "Gravity Belt Thickening and Lime Stabilization," *Water Environ. Technol.* (December 1990).
13. Westphal, P. and G.L. Christensen. "Lime Stabilization: Effectiveness of Two Process Modifications," *J. Water Pollut. Control Fed.* 55(11):1381–1386 (1983).

9 SLUDGE COMPOSTING

9.1 INTRODUCTION

In the past decade the number of sewage facilities that compost their sludge has increased steadily, for a number of reasons. Difficulties in permitting and developing long-term sites to apply liquid sludge have spurred interest in sludge composting. POTWs that rely on liquid application are often constrained by the weather for part of the year, and composting is an attractive backup method of sludge processing. In some areas landfill tipping fees are now high enough to make composting a financially viable alternative. A multipronged sludge management strategy has been adopted by most of the progressive treatment authorities, and many POTWs are choosing to compost part of their sludge stream. As a result, composting is an increasingly popular method of sludge processing. In 1991 there were 240 composting projects across the U.S., with 149 in operation and the rest either under construction, permitting, or design or closed down due to odor control problems. The technologies used by these projects are shown in Table 9-1.

Cocomposting sludge and yard waste is becoming a standard practice. Nearly one third of the composting facilities in operation today either report using yard waste as part of the compost mix or are planning to do so. Yard waste must be chipped and/or ground to reduce it to compostable size, and if the waste is mostly leaves and grass then the high nitrogen content may lead it to compost on its own. On the other hand, composting facilities occasionally charge a tipping fee for yard waste, so instead of buying amendment they are paid to take it, improving the ledger sheet.[1]

Odor control is the most common problem associated with composting. Composting smells, and the factors that impact odor generation include the type of sludge being composted, the mixing and aeration systems, compost stability, proximity to the nearest neighbor, and weather conditions. In some areas compost-related odors have resulted in the closing of composting facilities. Siting composting facilities is another common problem; a POTW without a site adjacent to the wastewater treatment plant may find that siting is a substantial obstacle. Some facilities have been constructed at the local landfill, especially when sludge is cocomposted with yard waste or wood waste.

Finished compost is much like well-rotted manure: rich, dark, and crumbly, with a musty, earthy smell. If a project generates a quality compost, experience has shown that distributing the end product is not a problem and that the demand may even

Table 9-1 Compost Projects in the United States, 1991

Technology	In operation	Under construction or permitting	Pilot plants	Not operating at this time
Static pile	2	1	—	—
Aerated static pile	76	13	6	3
Vermicomposting	1	—	—	—
Windrow	33	10	11	2
Aerated windrow	10	1	—	—
In-vessel	27	37	1	6

Source: Goldstein and Riggle.[21]

exceed the supply. Common markets include use within the public works system, distribution to local residents, landscapers, nurseries, golf courses, sod farms, and tree farms. In addition, large quantities of compost are being used as both daily and final cover on landfills.

For in-vessel or static aerated pile composted sludge to meet class A standards, the temperature of the sludge must be maintained at 55°C or higher for 3 days. For windrow composted sludge, the temperature of the sludge must be maintained at 55°C or higher for 15 days or longer. During this period, a minimum of five windrow turnings are required. Vector attraction reduction can be met by proper digestion, or by composting alone provided that the temperature is kept above 40°C for at least 14 days and the average temperature during this period exceeds 45°C. In addition to treatment, for a class A sludge product, either the density of fecal coliforms in the compost must be less than 1000 MPN/g total dry solids or the density of *Salmonella* sp. bacteria in the sludge must be less than 3 MPN/4 total dry solids. For class B sludge compost, using either in-vessel, static aerated pile, or windrow composting, the temperature of the sludge must be raised to 40°C or higher for 5 days. For 4 h during those 5 days, the temperature in the compost pile must exceed 55°C.

9.2 COMPOSTING PROCESSES

Composting is a simple, age-old process. The composting process breaks down the organic matter in sewage sludge into stable humus. As the sludge decomposes the piles emit methane, water, and heat. The carbon:nitrogen ratio is the critical factor in the composting process, and the nitrogen-rich sludge must be mixed with a carbon-rich amendment to compost successfully. Provided there is sufficient carbon, much of the nitrogen in the sludge is transformed to ammonia, which evaporates; hence, the resulting compost is considered to be a soil amendment rather than a fertilizer. Although the compost contains lower concentrations of phosphorus, nitrogen, and potassium than commercial fertilizer, the nutrient levels provided by most sludge composts are sufficient to meet the requirements of many types of plants. More importantly, the nutrients in sludge compost are present in an organic form which allows the gradual release of nutrient benefits. This slow release process satisfies the changing seasonal needs of a plant over a longer period of time. In addition to being a source of nutrients, compost improves soil quality by increasing the moisture and nutrient retention capacity of sandy soils and the drainage and aeration of heavy clay soils. Subsoil can be upgraded to topsoil by the incorporation of compost, making it an ideal amendment for disturbed land.

To compost sludge the dewatered cake is mixed with a bulking agent (typically wood chips) that acts as a carbon source and increases the porosity of the pile. A pile of the sludge/amendment mixture is constructed, and the sludge composts in a matter of weeks. Static pile composting is effective and inexpensive, but anaerobic pockets throughout the pile may generate substantial odors. Simple static piles are rarely used for that reason. Aerated static pile composting takes the same mixture and pile, but reduces the number of anaerobic pockets by blowing air through the piles. In windrow composting the sludge/amendment mixture is arranged in windrows which are turned a few times a week, while aerated windrow composting blows air through the windrows between turnings. These composting options are best suited to warm, dry climates. In-vessel composting provides the most completely aerobic composting environment, further reducing odor production and increasing the consistency of the sludge product. In-vessel systems require a significant capital investment and are best suited to cold, wet climates or highly populated areas. Table 9-2 provides a sketch of the advantages and disadvantages of the different composting processes.

9.2.1 Aerated Static Pile Composting

Static pile composting is the simplest method of composting, but is rarely used because of the potential for odor generation. The aerated static pile process is the most common method of composting sludge in the U.S. In aerated static pile composting, the aeration is facilitated by a series of perforated pipes running underneath each compost pile that are connected to a mechanical blower system. The pipes are covered with a layer of coarse bulking material which acts as a manifold to provide uniform aeration. Piles are finally covered with cured compost for insulation and to provide more uniform aeration. The air is blown or drawn through the piles with industrial fans or blowers. The pile is not turned and is therefore "static". Blower power typically ranges between 1 and 5 hp, depending on pile size, mixture density, and piping arrangements.[2]

The U.S. EPA requires that sludge which is composted using the static pile method be maintained at a temperature of at least 55°C for 3 days. Aerated composting is usually conducted for at least 21 days. Composting is performed on an impervious pad with a slight grade sloping to a leachate collection system. Prior to marketing, the compost is normally cured for approximately 30 days after the aeration period. Aeration and curing sometimes take place under a roofed structure or enclosed building.[2]

Coeur d'Alene, ID provides an example of a typical aerated static pile composting facility. The materials balance of this composting facility is provided in Table 9-3. There are four types of equipment used in Coeur d'Alene: a mixer, a front-end loader, blowers, and a screen. The front-end loader is used to load the mixer, move the raw mix to compost piles, break down the piles, load the screen, move the screened product to cure, and load trucks; the mixer blends the sludge and amendment, the blowers aerate the compost piles, and the screen is used to separate the finished compost from the remaining amendment.

Table 9-2 Advantages and Disadvantages of Composting Processes

Aerated Static Pile Systems

Advantages

- Moderate capital cost. Required capital equipment consists of perforated piping (usually plastic), concrete pads, low-pressure fans and duct work, front-end loaders, a screen, temperature probes, and a condensate/leachate collection system. Roofed or enclosed structures will increase costs and may not be necessary depending on site-specific conditions.
- High pathogen destruction. Uniform aeration and pile insulation help maintain high temperatures required to kill the organisms.
- Good odor control. Pile insulation, combined with uniform aeration with blowers in the suction mode, allows odors to be treated as a point source.
- Good product stabilization due to efficient aeration and maintenance of temperature at optimum levels.

Disadvantages

- Requires more land than in-vessel systems.
- Rain or snow may hamper the process if uncovered and may result in a less uniform product. Cold will not affect the system, but may affect operators and equipment.
- Depending on site selection, expensive odor control equipment may be required.

Windrow Systems

Advantages

- Rapid drying due to moisture released when turning windrow.
- Drier product allows easier separation and high recovery of bulking agent (if practiced).
- Capacity to handle high volumes of sludge.
- Good product stabilization.
- Low capital cost if facility is not covered.

Disadvantages

- Large space requirements — space is needed between windrows and they cannot be piled high (more than 7 ft) or wide (more than 18 ft) due to machinery limitations and aeration requirements.
- High equipment operation and maintenance costs due to frequent turning of the piles unless a specialized compost-turning machine is available.
- Requires intensive monitoring of temperature (more than static pile) due to the reliance on operator skill and consistency in turning windrow piles.
- Turning piles may release localized odors.
- Inability to operate when rainfall occurs unless the piles are covered.
- Large volume of bulking agents needed to enhance aeration.

In-Vessel Systems

Advantages

- Low land requirements.
- Better process control than outdoor systems.
- Not affected by adverse weather conditions.
- Enclosed system allows easy addition of odor control equipment.
- Potential heat recovery can assist in maintaining optimum composting temperature.
- Reduced operating labor requirements.

Disadvantages

- High capital cost, highly mechanized system, and equipment of relatively new design for the U.S.
- Lack of operating data for large-volume systems.
- Operation relies on specialized mechanical equipment. Possible long downtime and high maintenance costs in the case of component failure.
- High maintenance requirements, inherent with a fully mechanized system.
- The potential for incomplete product stabilization due to short reactor detention times.
- Less operational flexibility, i.e., airflow variance, compost volume variance.
- Odors have been a problem with some systems.
- Some systems have been plagued by materials conveyance problems and lack of operating reliability.

Source: Burnett.[2]

Table 9-3 Aerated Static Pile Materials Balance, Coeur d'Alene, Idaho

Item	Volume (yd³/day)	Total weight (lb)	Dry weight (lb)	Volatile solids (lb)	Bulk density (lb/yd)³	Solids content (%)	Volatile solids (%)
Sludge	14	23,104	5,314	2,710	1,600	23	51
New wood chips	9	5,198	3,119	2,963	600	60	95
Recycle	35	24,192	13,306	10,644	700	55	80
Mix	55	52,494	21,739	16,317	950	41	75
Loss			1,631	1,631			
Unscreened	46	36,560	20,108	14,686	800	55	73
Recycle	35	24,192	13,306	10,644	700	55	80
Compost	12	12,368	6,802	4,042	1,000	55	59

Source: Bennett, Epstein and Porter.[3]
Note: Wood chip recovery efficiency at 80%.

The sludge is trucked from the treatment plant to the compost facility, where it is mixed with wood chips at a 2.5:1 volumetric ratio of wood chips to sludge. The goal is to have a mix of 40% solids, blending a sludge cake of 23% solids with wood chips at 60% solids.

Mixing is critical to create a porous sludge/amendment mixture that will allow air to pass through the pile. The trailer-mounted, self-contained, diesel-powered mixer has four augers inside the mix box and a discharge auger arrangement. Wood chips and sludge cake are loaded into the mix box, which then operates for about 5 or 10 min, at which time sludge is discharged from the side of the mix box.

The piles are constructed on top of a 12-in. layer of wood chips which covers the 4-in. PVC pipes attached to the blower. The front-end loader piles the sludge mixture on the bed of wood chips until the mound is about 5 ft high. To complete the pile a cover of unscreened compost is used as an insulation blanket; it also acts as a biofilter to control odor emissions.

The facility uses ten 2-hp blowers, each of which is connected to two perforated PVC pipes aerating one compost pile. The blowers can either suck air through the pile in vacuum mode, so the air can be discharged to the odor control system, or blow air through the piles in pressure mode, in which case the air goes directly into the environment. Odor control is provided by a biofilter that uses compost to filter the exhaust air. It also provides a medium in which bacteria that break down odor components can grow.

The composting process takes about 21 days, during which the piles are aerated continuously. The compost piles are then dismantled. If the compost has reached 55 to 60% solids, the material is picked up by the front-end loader and deposited into the feed hopper of the screen. If the compost is too wet to handle it is restacked over perforated pipes for further drying. The screened compost is cured for 30 days, after which the product is ready for distribution.

During the first year of operation much of the compost was used by the city for beautification projects and trials. This generated interest in the product by local nurseries and other horticultural enterprises and introduced the general public to the benefits of sludge compost. The following year Coeur d'Alene expanded their composting operation and began bagging their compost for local distribution.[3]

9.2.2 Vermicomposting

Vermicomposting is a Californian twist on the conventional static pile. The Fallbrook Sanitary District of Fallbrook, CA pioneered this process in the U.S. with a pilot project in the early 1980s. The District found that there was a substantially greater demand for the vermicompost than for its static pile compost and initiated an ambitious full-scale project in 1986.[4]

Conventional composting relies on microorganisms to stabilize the sludge, while vermicomposting uses the common earthworm, a higher order organism, to accelerate the stabilization of organic material. Earthworm excreta, or castings, are an excellent soil conditioner with high water retention capacities, and vermicompost is an attractive sludge product with superior growth-enhancement properties.

Vermicomposting is accomplished in two stages. The temperatures in vermicomposting are never high enough to comply with EPA requirements, so the sludge must be pretreated by static pile composting to facilitate pathogen reduction. After about 30 days in a static pile, the compost is transferred to the worm beds for an additional 60 to 90 days of vermicomposting. The worms used are brandling worms *(Eisenia foetida)* and red worms *(Lumbricus rubellus).*

The District pumps anaerobically digested sludge at 1 to 1.3% solids 0.5 mile to 12 concrete drying beds of 20 × 165 ft, where the sludge stays until it is 15 to 18% solids — about 30 days in summer and 60 days in winter. The dried sludge is transported to the composting facility, and straw is used as the amendment. The straw is mixed with the sludge at a 2:1 or 3:1 ratio, depending on the solids content of the sludge. After 30 days in a static pile, the material is screened and transferred to a vermicompost bed for further processing.

The vermicomposting beds are about 8 ft wide and of varying lengths, arranged on an open tarmac. The static pile compost is mixed with straw and layered 3 in. thick on the bed; then it is seeded with about 1 lb of worms per cubic foot of compost. The worms require an initial acclimation period of about 1 month to become accustomed to their new environment, after which 4 to 6 in. of sludge compost is applied to the top of the beds every week. About once a month additional straw is added to maintain porosity. In the hot southern Californian sun it is necessary to water the beds regularly; the District parallels a sprinkler system with the worm beds. The amount of water used is dependent on the weather, and water use can be reduced by shielding the beds from direct sunlight.

The worms feed on the newly applied compost at the upper levels of the bed, depositing their castings in the lower levels. Sludge compost is applied continuously to the top of the bed until it is 3 ft high, a process that takes about 6 months. The worms are concentrated in the top 6 to 8 in. of the bed; the rest of the bed is stabilized vermicompost. The vermicompost is transferred to the screening facility for further processing; the top section containing the worms is returned to the original bed, where it serves as the seed for the next generation. Compost and straw are reapplied and the cycle repeats itself. The earthworm population doubling time appears to be 60 to 90 days. The process is easily expanded by partitioning the top layer into two parts and reseeding two beds.

Vermicomposting is substantially more complex than conventional composting techniques. There are greater handling requirements, and it uses more amendment.

It also uses about four times the area of static pile composting and takes twice the process time. It does not raise the temperature of the sludge, so pretreatment (static pile composting) is required to meet the EPA guidelines. The process must be monitored closely to ensure that the pile conditions create an optimal environment for the worms. Moisture, pH, temperature, and feed levels must be carefully maintained or the worms may die, effectively shutting down the process. Nonetheless, the quality and marketability of the final product are so clearly superior to traditional composted sludge that it is readily acceptable to a wide range of customers, from the home gardener to the commercial cut-flower grower. Demand for the material has far outstripped the supply. This is one sludge disposal alternative that is readily accepted and even encouraged by the local population.[5]

9.2.3 Windrow Composting

In windrow composting the pile is regularly turned for aeration, providing better mixing and quicker composting. As compared to the static pile process, windrow composting is more reliably aerobic, less likely to have odor problems, and far more equipment intensive.

There are two types of windrow composting in use today: conventional and aerated. The least complicated is the conventional windrow, which relies on natural aeration rather than forced mechanical aeration. Long parallel rows of a mixture of sludge cake and amendment are constructed, and the piles are turned frequently with mobile equipment during a period of several weeks. Aerated windrows, combining aspects of aerated static pile and conventional windrow processes, are gaining popularity. Long windrows are constructed over an aeration trough that is aerated with mechanical blowers and turned using mobile machines.

For unrestricted use of the compost, the U.S. EPA requires the mixture to attain a temperature of 55°C or greater for 15 days during the composting period. During the 15 days the windrows must be turned at least five times. Most windrow operations compost for 4 to 6 weeks, after which the composted sludge is stockpiled to cure for an additional 4 to 6 weeks before distribution.[2]

The conventional and aerated windrow processes, shown in Figure 9-1, are both based on the construction of a windrow from a mixture of sludge cake, recycled finished compost, and amendment. The amendment (sometimes called a bulking agent) is a material that is mixed with the sludge cake to provide windrow porosity, proper moisture content, and structural stability while acting as a supplemental food source for microorganisms. The windrows are turned with mobile equipment and composted for a period of several weeks. The finished sludge product is dry, disinfected, stabilized, and nonodorous — in short, suitable for distribution and marketing. The aerated method relies on induced aeration as well as turning to supply the requisite oxygen.

Windrows can be built in a number of different ways. Cake and amendment can be loaded together into a tractor trailer that dumps the loads end-to-end to form long rows. These rows are mixed with a composting machine that straddles the windrows or with a front-end loader. Alternatively, the sludge cake can be laid on the field in rows with amendment in adjoining parallel rows, with a front-end loader being used to combine the two.

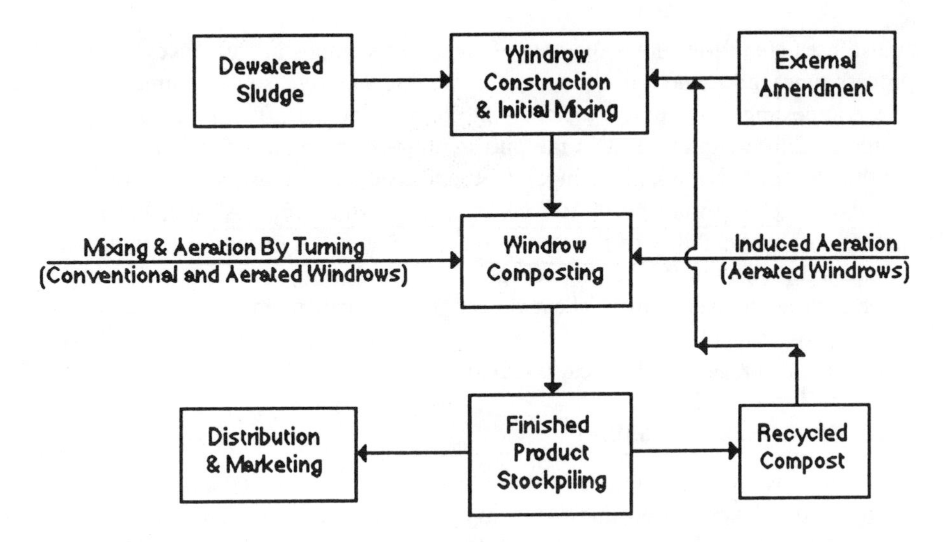

Figure 9-1 Windrow and aerated windrow composting processes. (From Hay, J. C. and Kychenrither, R. D., *J. Environ. Eng.*, 116(4), 746, 1990. With permission.)

Windrow size is an important composting variable. Windrows are usually 4 to 5 ft high, about 14 ft wide, and hundreds of yards long. For the composting process to work, the heat created by the thermophilic decomposition processes must be greater than the heat lost through the windrow surface. By increasing the cross-sectional size of the windrow, the surface to volume ratio decreases and heat losses go down. Studies in California showed that for each 0.5 m^2 (5 ft^2) increase in cross-sectional area, the internal temperature of the pile increased by 1.2°C (2°F).[6] Windrow temperatures increase with the size of the windrows, which should be built as large as the composting equipment allows.

For the windrow process to work, the cake and amendment must be thoroughly mixed. Composting machines straddle the windrow and use a high-speed rotating drum fitted with fixed teeth or flails to thoroughly mix the dewatered cake and amendment. After construction, a windrow requires from 30 to 50 days to complete the composting cycle, although rainy weather can prolong the process by many weeks.

The rate of decomposition of the organic matter is dependent on the moisture content, and the optimal moisture level is in the range of 50 to 60%. Sludge is, of course, quite wet, and the sludge should be dewatered as much as is economically feasible and mixed with a dry amendment or recycled product. If the moisture level exceeds 60%, the sludge mixture will not have the structural integrity to maintain shape and the windrow will slump. Below 60% the porosity is measurable and air will be able to penetrate the pile; below 50% moisture the biological activity will be reduced and the composting process slowed.

The frequency that the windrows are turned has a marked influence on the composting process,[6] as seen in Figure 9-2. Turning mixes the contents of the windrows, lightens the pile (increasing porosity and promoting aerobic conditions), releases water vapor, and exposes all of the material in the windrow to the high interior temperatures.

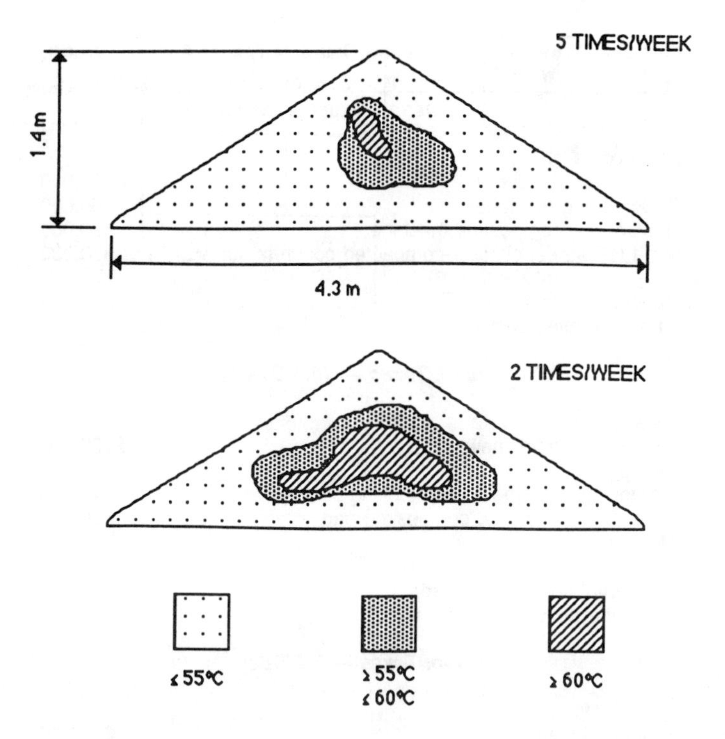

Figure 9-2 Turning frequency and pile temperatures for windrow composting. (Redrawn from Reference 23.)

Table 9-4 provides the costs associated with four Texas windrow composting facilities.

9.2.4 Aerated Windrow Composting

In aerated windrow composting, forced mechanical aeration supplements the aeration provided by turning with mobile equipment. Aerated windrows require less land, smell less, dry more quickly, offer better process control, and are less sensitive to rainy weather. Aerated windrows are also much more capital intensive, requiring the installation of aeration odor control facilities.[7] Figure 9-3 is a schematic of the aerated windrow facility.[7] The blower sits on top of a pressurized sump and typically operates at a pressure of 18 to 24 in. (46 to 61 cm) of water column. The sump, which separates the leachate and condensate from the composting process, is connected to an aeration channel by a pipe. The interface between the air channel and the windrow is shown in detail. A perforated steel plate induces sufficient head loss to ensure that the air is spread across the length of the interface. Next is a geotextile mat covered with a screen, which ensures further air distribution. The final layer of sand keeps the compost from plugging up the interface. This sand must be replaced about every 6 months.

The air drawn through the compost pile ensures that the pile is aerobic, and it also accelerates the drying process. In conventional windrow composting the surface of the pile is dried by solar evaporation and the interior is dried by the natural ventilation of the pile. Aerated windrows increase the interior airflow, providing additional

Table 9-4 Windrow Composting Costs in Texas: Austin, Laredo, Bryan, and Temple

Austin, Texas — 15 DT/Day	
Capital costs	
2 Front-end loaders	$400,000
1 Screening machine	60,000
1 SCARAB composting machine	170,000
4 ha (10 acres) of roller-compacted concrete surface	970,000
Total capital costs	$1,600,000
Annual operating costs	$23/DT
Laredo, Texas — 10.4 DT/day	
Capital costs	
Front-end loader (new)	$150,000
Dump truck (new)	55,000
Screening equipment and stacking conveyor	40,000
2 ha (5 acres) paved composting pad	590,000
Site improvements	25,000
Total capital costs (estimated)	$860,000
Annual operating costs (estimated)	$34/DT
Bryan, Texas — 7 DT/day	
Capital costs	
Front-end loader	$54,800
Compost screening equipment	95,700
Tub grinder	200,000
2 ha (5 acres) gravel pad	116,000
Total operating costs (estimated)	$621,500
Annual operating costs (estimated)	$30/DT
Temple, Texas — 6 DT/day	
Capital costs	
Rubber-tired loader	$75,000
SCARAB composting machine	185,000
Tub grinder	200,000
0.36 ha (0.9 acre) covered slab	240,000
0.6 ha paved covered composting pad	255,000
Total capital costs	$900,000
Annual operating costs	$44/DT

Source: Burnett.[2]
Note: DT = dry tons.

moisture removal. However, if a windrow is overaerated the pile cools and biological activity slows down. Routine monitoring of oxygen and temperature inside the windrow is used for aeration control.

The fixed aeration system of a windrow composting facility is a major capital investment. The resulting sludge compost is drier than compost from nonaerated windrows, and there are fewer odors generated during the composting process.

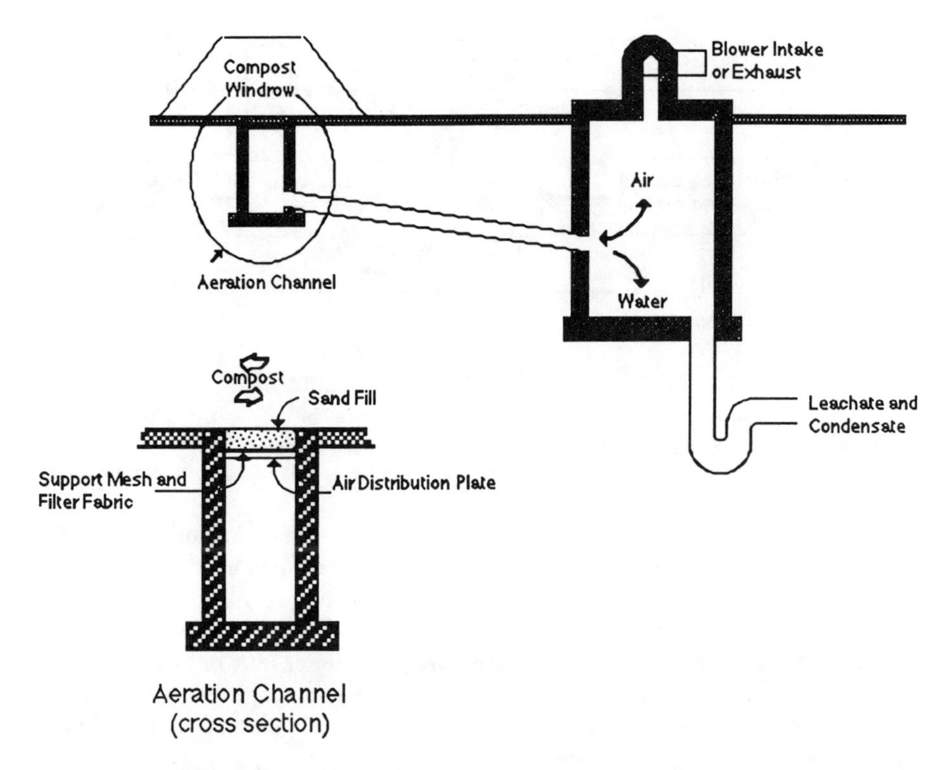

Figure 9-3 Schematic of an aerated windrow composting facility. Redrawn from Hay, J. C. and Kuchenrither, R. D., *J. Environ. Eng.,* 116(4), 746, 1990. With permission.)

9.2.5 In-Vessel Composting

In-vessel composting offers several advantages over static pile or windrow composting: there is increased control over the composting environment, so the sludge product is more consistent; there is little land required; there is better public acceptance for enclosed facilities; there is less labor required; and the odors are more easily contained and controlled.[8]

An in-vessel composting facility is composed of a number of related components: materials, materials handling, reactors, aeration equipment, odor control facilities, exterior curing and storage facilities, and compost distribution and marketing. Sludge cake, recycle, and amendment are mixed and placed in the aerator reactors for composting. The reactor is a closed vessel which may be circular or rectangular, constructed of structural steel or reinforced concrete. Air is injected to control the temperature, remove the moisture, and promote biological activity. After passing through the compost, the air is exhausted to the treatment system and dispersed into the atmosphere. When the composting is completed the material is removed from the reactor for curing and storage and, ultimately, distribution. The in-vessel composting process is shown in Figure 9-4.

The in-vessel reactor has two stages: active compost and curing. The first stage is characterized by high oxygen uptake rates, high temperatures, rapid degradation

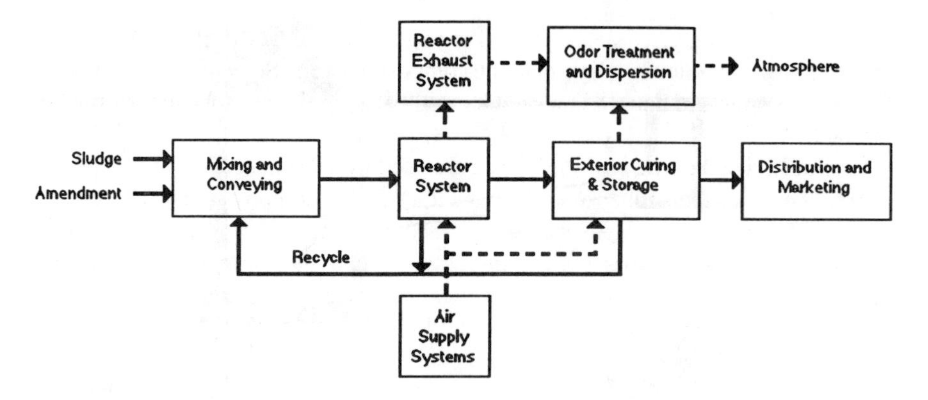

Figure 9-4 The In-vessel composting process. (Redrawn from Johnston, J. R., Donovan, J. F., and Pincince, A. B., *Water Environ. and Technol.,* May 1990. With permission.)

of biodegradable volatile solids, and a high potential for odor production. The second stage is characterized by lower temperatures, decreased oxygen uptake rates, and a lower, but still significant, potential for odor production. No precise distinction exists between the two stages. At in-vessel facilities, the first stage is performed in a reactor. The second stage can be performed in a reactor, an exterior pile, or both.[9]

There are two classes of reactors: plug-flow reactors and agitated bed reactors. The primary differences between these reactors are the configurations of the aeration system and the discharge mechanisms.

In a vertical plug-flow reactor, the mix is placed on the top of the reactor and moved as a plug to the bottom, where it is discharged by a rotating screw. In a horizontal plug-flow reactor, the mix is introduced at one end of the tunnel and travels the length of the tunnel as a plug. The material is propelled by a mechanism such as a moving floor or hydraulically operated door.[10]

The agitating bin process combines features of windrow and static pile systems within a building. Composting in this system is carried out in concrete bays; the compost is moved along the bay with an automated machine which agitates the mixture to maintain uniform aeration. Aeration is also supplied by a blower and a perforated pipe system embedded beneath the bins. Compost is fed to the receiving end of the bays in batches or plugs with conveyors or front-end loaders, depending on the particular design. Each plug stays in the system between 21 and 30 days. This technology is relatively new to the U.S., but has had about 15 years of full-scale use in Japan.[2]

When the community of Palmetto Dunes on Hilton Head Island, SC was faced with rising costs for sludge disposal coupled with a very small sludge processing site that was located in an environmentally sensitive area, it opted to construct a totally enclosed, modular in-vessel composting system. Two Dynatherm® in-vessel reactors, manufactured by Compost Systems of Cincinnati, OH, were factory assembled and shipped to the job site by truck in November 1990. Cranes set the reactors in place on previously prepared steel supports. The entire composting system was erected in less than 2 weeks after delivery of the reactors.

The facility is designed to treat from 35 to 45 tons of sludge per week, depending on the dry solids content of the dewatered cake. Sludge from the secondary clarifiers is pumped into a gravity thickening tank and then to a 1-m belt filter press for dewatering. The dewatered sludge, bulking agent, and recycled compost are stored in a building immediately adjacent to the compost reactors. Screening equipment and finished compost are housed in a shed located between the reactors and the thickening tank.[11]

Sawdust, wood chips, and yard waste are used as bulking agents. The yard waste is processed through a tub grinder to make it a suitable composting amendment, and it breaks down almost completely during the composting process. Tables 9-5 and 9-6 provide materials balances for the in-vessel system. Palmetto Dunes is a resort community, so the sludge production varies from season to season. Detention time in the reactors varies from a minimum of 14 days under peak conditions to as long as 28 days in off-peak periods.

Several times each week, dewatered sludge is combined with recycled compost and yard waste and placed in a batch mixer. The ratios are calculated by choosing mixing ration that ensure that the dry solids content of the combined materials stays above 40%. A small front-end loader with a 0.5-yd^3 bucket measures the materials and transports them from the storage building to the batch mixer. When all of the materials are placed in the mixer, the unit goes through an automatic cycle and then discharges the mix to a screw conveyor and bucket elevator that carries it to the top composting reactor.[11]

When loading is complete, temperature probes are inserted into the composting mass. These probes allow a computerized process controller to regulate the blowers which provide air to the reactors. The reactors have four separate aeration zones, each equipped with a 3-hp blower rated at 600 ft^3/min. During the composting cycle, temperatures in excess of 58°C(136°F) are reached for a minimum of 3 days, exceeding EPA requirements.[11]

The reactors use a "moving floor" similar to the devices used in truck trailers to transfer the composting mass from the feed to the discharge end of each reactor. After 7 to 14 days of detention, the composting mass passes from the top reactor through a mixer and into the bottom reactor, where it stays for another 7 to 14 days. The remixing ensures that a porous compost mass is maintained and that the bioculture is exposed to new food sources. When compost is discharged from the reactors, it either goes to a curing area or is recycled as part of a new mix.[11]

Cost for installation of the entire system — reactors, odor control, conveyors, mixer, and storage building — was about $600,000. Composting operations and maintenance require less than 20 hours per week, and use of yard waste as a bulking agent has reduced operating expenses. As a result, the total life cycle costs on a per ton basis of composted sludge have been less than $50/ton.[11]

The costs for in-vessel composting cover quite a range. Clayton County, GA had in-vessel composting costs of $73/ton in 1989.[12] Hamilton!, OH (the city officials legally added an exclamation point to the name) had an in-vessel composting system that cost $112/ton in 1989.[13] Cape May County, NJ has an in-vessel system that costs $223/ton.[2] The costs projected in Burlington, VT for in-vessel composting were over $400/ton.

9.3 ODOR CONTROL

There are a number of potential sites for odor generation during the composting process. The dewatering process, the cake storage area, the mixers and composting

Table 9-5 In-Vessel Composting with Sawdust and Wood Chips, Hilton Head Island, SC

Materials	Bulk weight (lb/ft³)	Dry solids (%)	(ton/week)	yd³/week
Sludge	60	16	35	43
Sawdust	18–22	55–65	6–7	21–31
Wood chips	20–24	55–60	13–15	42–45
Recycled compost	32–35	54–58	31–34	70–75
Compost mix	37–40	40–42	85–90	160–165
Compost discharged	32–35	54–58	55–60	125–130
Less compost recycled	—	—	(31–35)	(70–75)
Finished compost	32	55+	24–26	55

Source: Rutland and Ryan.[11]

Table 9-6 In-Vessel Composting with Yard Waste, Hilton Head Island, SC

Materials	Bulk weight (lb/ft³)	Dry solids (%)	(ton/week)	(yd³/week)
Sludge	60	16	35	43
Yard waste	18–22	55–65	25–28	86–90
Recycled compost	30–35	54–58	23–25	55–60
Compost mix	36–39	40–42	82–86	165–170
Compost discharged	30–34	54–58	54–59	126–131
Less ompost recycled	—	—	(23–25)	(55–60)
Finished compost	32	55+	31–34	71

Source: Hay and Kuchenrither.[6]

piles, the curing piles, and the conveyance system used to transfer the sludge between all of these areas are all potential sites for odor generation.

Odor is a subjective parameter, and the range of sensitivity to odor in a given population varies widely. Odor measurement typically uses a ten-person panel, and the most common parameter for detecting odor is ED 50, the number of fresh air dilutions required to reduce the odor level of a sample so that only 50% of the panel can smell it. Unfortunately, if the ED 50 is used as the odor control parameter, it is conceivable that one half of the population would smell the compost facility. This problem has led to the creation of ED 5, the level at which only 5% of the population may perceive composting odors.[14]

It is often said that odor isn't a problem until the neighbors start complaining. However, POTWs are generally located near population concentrations, and once people have become sensitized to composting odors it may be impossible to keep running the facility unless odor reduction is effective. Optimizing the composting process will reduce the production of most of the offensive compounds that are generated during composting. Exhaust air odor control is usually a key component of a successful composting facility.

Odorous emissions must be dispersed and diluted so that concentrations reaching people are below detection thresholds. Ideally, the dispersion characteristics of a proposed site should be modeled during project planning. Dispersion modeling should consider the effects of local meteorology, including seasonal and diurnal variations and worst-case weather conditions such as inversions, topography, the strength and location of the odorous discharges, and the location and proximity of the neighbors. Results can be used to screen sites with poor dispersion characteristics.[15]

9.3.1 Malodorous Compounds

Composting is a smelly process. In particular, fatty acids, amines, aromatics, inorganic and organic sulfur, and terpenes all smell, and all are important to composting.[14]

The fatty acids include acetic, propionic, and butyric acids, which are produced under anaerobic conditions. As long as the compost pile does not begin to go sour and anaerobic, these compounds do not appear to be much of a problem. The amine family includes such aptly named compounds as putresine and cadaverine. These smelly products of microbial decomposition are a by-product of the breakdown of amino acids, which is how *Escherichia coli* metabolizes. Amines are formed at the lower pH range during anaerobic fermentation. Aromatics are present in influent wastewater and can be absorbed into the sludge that ends up at the composting plant and subsequently volatilized and released during aeration. Aromatics are also produced by the breakdown of lignins (found in wood chips) during the aerobic composting process.

The inorganic sulfur compound hydrogen sulfide is commonly encountered throughout a POTW, and it is produced through several anaerobic pathways. The sludge may be producing hydrogen sulfide when it arrives at the site. Once the sludge has been mixed with amendment and is built into a suitably aerobic pile, hydrogen sulfide emissions are very low. However, if the sludge and amendment are poorly mixed or if the mix is too wet, there can be clumps of material that become anaerobic, and hydrogen sulfide forms in these clumps. Combining carbohydrates with sulfate produces hydrogen sulfide, and the breakdown of amino acids (which are metabolized by *E. coli*) produces inorganic sulfur and hydrogen sulfide as end products.

Organic sulfurs are the worst offenders. The ED 50 values for dimethyl sulfide and dimethyl disulfide are about 3 and 2 ppb, respectively, and these compounds can be detected at levels between 0 and 10,000 ppb at a composting facility. They degrade during sample analysis and are difficult to quantify, but it is easy to identify their presence by nose: it's just the quantification and speciation that is problematic.

Organic sulfur compounds come in a trio that can be produced and exchanged with one another. Methyl mercaptan, produced under both anaerobic and aerobic conditions, can be oxidized to produce dimethyl disulfide (another stinker), which breaks down to dimethyl sulfide aerobically or methyl mercaptan anaerobically. The net effect is that there are three organic sulfur compounds present in composting which vary in quantity with the daily conditions in the pile. It's likely on any given day that each of the three compounds is being formed and degraded somewhere on the pile, whether or not they are detected.

Terpenes are products of wood and will be present at any composting facility that uses sawdust or wood chips as an amendment. Terpenes can be found at concentrations of 0 to 500 ppb, and the odor threshold is 6 ppb. These compounds are very stable and difficult to remove, although recent efforts with a mist-type scrubber have been promising.

9.3.2 Odor Control by Process Optimization

Composting is a two-step operation of composting and drying. The two steps are directly related: the composting reaction —the degradation of volatile solids — provides the heat for pathogen destruction and evaporation. The composting process can be optimized to minimize production of malodorous compounds, but optimization depends in part on the specifics of an individual composting operation. Experience has shown that the strongest compost odors occur within 1 to 2 weeks after initiation of composting at temperatures greater than 50°C.[15]

To reduce odors during composting, the pile conditions should be as perfectly aerobic as possible. The sludge and amendment should be well mixed to reduce the number of anaerobic clumps in the mixture. Similarly, the pile should be well aerated to provide a steady source of oxygen for the biological activity. Some composting operations aerate with pure oxygen to reduce odors.

Air serves to remove moisture, act as a source of oxygen, and provide a method of temperature control. At the same time, a higher air flow through the pile creates a larger volume of air to be deodorized before it exits the composting building. Some compost facilities maintain a steady airflow throughout the composting process; others cycle the blowers on and off to maintain the correct pile temperature. During active composting the oxygen level in the pile decreases rapidly when the blower is off, as shown in Figure 9-5.

Moisture is another important odor parameter. The more moisture, the more odor. Moisture can be reduced by dewatering the sludge to a higher solids content and by increasing the amount of amendment added to the mix. Decreasing the moisture content of the mix means that less air is required to dry and stabilize the mix, decreasing odor production.

Composting temperatures are quite important to odor generation, with the smell increasing as the temperature rises (Figure 9-6). If the sludge/amendment mix is low in volatile solids then little heat will be generated, evaporation will be minimal, and the potential for odor generation increases.

There seems to be a direct correlation between the oxygen demand of the sludge and odor production. Secondary sludge is more acclimated microbiologically and is more readily degraded than primary sludge. Wilber and Murray[14] found that secondary sludge smelled the most during composting, followed by combining secondary and primary sludge, primary sludge, and, finally, anaerobically digested and nitrified sludge.

9.3.3 Amendment Choice

Amendments used in sludge composting include recycled finished compost, wood chips, sawdust, rice hulls, ground straw, agricultural waste, and yard waste. The ideal amendment has a high solids content, high volatile solids, low moisture content, and a low bulk density. An amendment with better bulking properties creates a lighter, drier mix. The porosity of the pile promotes oxygen penetration and increases the supply. As a result, an amendment with a low bulk density, such as rice hulls or ground straw, produces higher compost temperatures than sludge amended with recycled finished compost alone.[16]

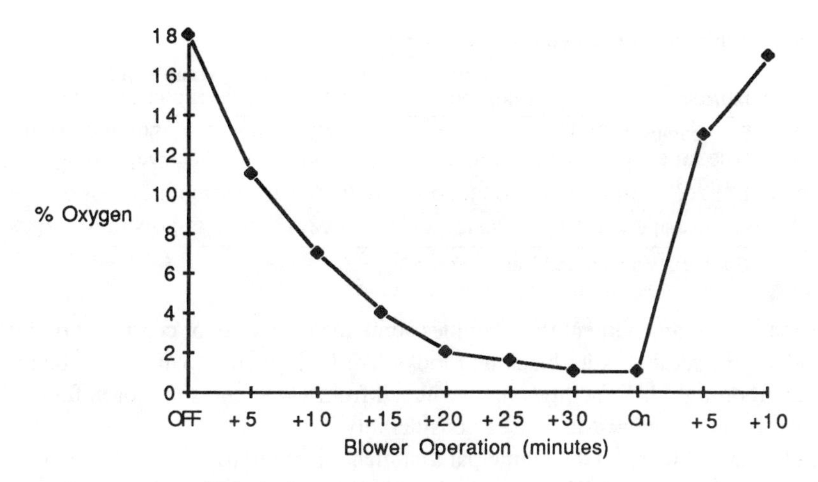

Figure 9-5 Oxygen depletion and regeneration. (From Murray, C. M. and Thompson, J. L., *Biocycle,* July 1986. With permission.)

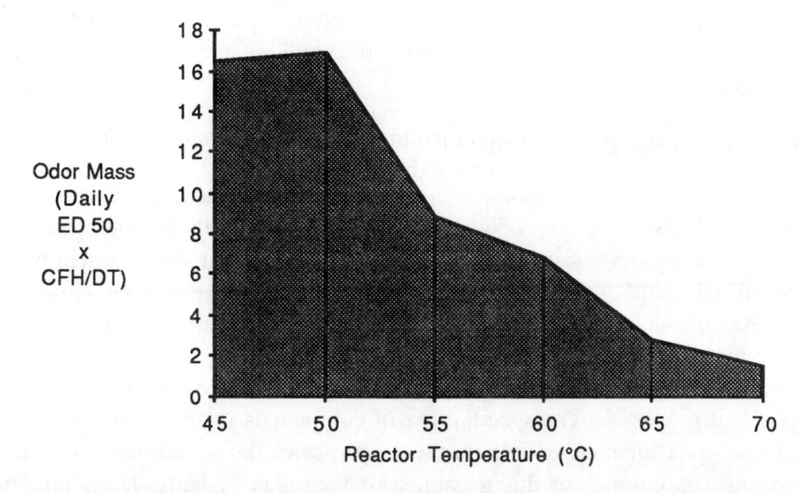

Figure 9-6 Odor generation decreases as composting temperatures rise. (From Wilber, C. and Murray, C., *Biocycle,* March 1990. With permission.)

The choice of amendment can affect the odor generation of a compost pile. An amendment with a higher solids content makes a mix with less moisture, and an amendment with lower bulk density provides porosity and improves airflow through the pile. Properties of various amendments are shown in Table 9-7.

The fineness of the amendment can affect the marketability of the finished compost. Greenwich, CT facility personnel found that a significant amount of their compost was used as surface dressing on new lawns. When wood chips were used as an amendment the chips became visible after rainstorms, a characteristic that homeowners found unattractive. By switching to yard waste as an amendment the operators noticed increased interest in their compost. South Portland, ME saw a similar increase in interest when it switched from wood chips to sawdust.[17]

Table 9-7 Properties of Amendments

Material	Bulk density (kg/m^3)	Total solids (%)	Volatile solids (%)
Bulk compost	590	63	50
Rice hulls	130	74	95
Sawdust	260	62	96
Straw (ground)	224	73	80
Cotton-gin waste	249	74	69

Source: Kaneko and Yamanaka.[19]

Bioash is an amendment that provides some measure of odor control; it produces a sludge product that is jet black and looks like high-quality loam. Kennebunkport and Scarborough, ME facilities that switched from wood chips to bioash found that interest in their compost increased considerably.[17]

In Coeur d'Alene, ID the municipal treatment plant ran trials to evaluate the most efficient bulking agent. The amendments tested include cedar chips, yellow pine wood chips, seed hulls, sawdust, and red fir wood chips. Cedar chips were found to be too wet to compost well; yellow pine chips were too expensive and often wet; seed hulls were expensive and caused odors when used as the sole bulking agent; red fir wood chips, when dry, were the most economical bulking agent, although they have a low recycling rate.[3]

9.3.4 Exhaust Air Odor Control

Airflow serves to remove moisture, act as a source of oxygen, and control the pile temperature. It also transports odors. The first step in controlling composting odors is to carry out composting in a closed structure. The simplest structure will prevent the prevailing winds from wafting the smell to the nearest receptors. However, a higher level of odor control is generally required, and exhaust air odor control is a standard feature of most composting facilities.

Compost odors are caused by a variety of compounds that differ in their chemical and physical properties. The specific mix of compounds in an odorous gas stream will change over time, depending on the properties of the mixed materials and the composting conditions. For this reason, odor treatment systems should employ a broad spectrum of removal mechanisms to be fully effective. The range of available odor control methods is shown in Table 9-8.

Chemical wet scrubbers remove odorous compounds from the airstream through the adsorptive and/or oxidative capacity of the scrubbant solutions. There are two commonly used types of wet scrubbers: mist and packed towers. A mist scrubber achieves contact between the odorous compounds and the scrubbant solution by breaking the scrubbant into atomized droplets that are dispersed through the airstream. In packed tower systems the liquid is divided into a number of slow-moving films that flow over a packing media. The air being treated is typically introduced at the bottom of the vessel and flows up through the media. The state of the art in wet scrubbing consists of a minimum of two stages; three stage systems are not unusual.[18]

Treatment systems that rely on only one or two removal systems, such as wet scrubbers with either single- or two-stage packed towers and ozone oxidation chambers, have been somewhat unsuccessful in controlling composting odors. Odor systems

Table 9-8 Odor Control Methods

Odor Control Methods
Water scrubbing
Water scrubbing plus chemical scrubbing
Water scrubbing plus ozonation
Water scrubbing plus ozonation plus adsorption
Chemical scrubbing
Chemical scrubbing plus adsorption
Chemical scrubbing plus ozonation
Chemical scrubbing plus ozonation plus adsorption
Adsorption
Ion exchange resin plus adsorption
Soil treatment
Aeration
Ozonation
Perpendicular combustion

Source: Kaneko and Yamanaka.[19]

that use multiple-stage wet scrubbers employing sulfuric acid, sodium hypochlorite, and sodium hydroxide solutions are generally more effective. Cooling the exhaust air by dilution with cooler air before scrubbing improves odor control by making the volatile nitrogen compounds more soluble in the acidic water of the scrubber.[8]

Activated carbon units typically consist of a cylindrical vessel which contains two beds of granular activated carbon. The odorous air enters an open cavity within the vessel and is distributed across the face of each bed. As the airstream passes through the activated carbon, odorous constituents are removed. Activated carbon or impregnated charcoal is often used to remove the medium and higher molecular weight malodorous compounds.[18]

Activated carbon is not widely used for composting facilities. The adsorptive capacity of the carbon is quickly exhausted, and regeneration or replacement is costly and can take several days. Facilities which use activated carbon need to have a standby unit as well, increasing the costs of odor control.

Bubbling odorous gases through treated wastewater has been used successfully to control odors at a number of sites around the country, as have soil and compost biofilters. Biofiltration consists of directing the exhaust stream into a series of perforated pipes laid in a bed of gravel and covered with an organic medium. As the airstream filters up through the medium, odorous compounds are removed by a combination of physical, chemical, and biological mechanisms. Biofilters can be constructed for moderate cost when compared to other odor control technologies, although care must be exercised in selecting an appropriate gas detention time and superficial velocity.

In the past decade Japan has developed a number of biological deodorization methods which use microorganisms to eliminate odors. The soil treatment method, a carefully engineered soil biofilter, uses microorganisms in the soil to adsorb and degrade odorous compounds. The soil is black dirt mixed with chicken manure and compost and is formed into a layer about 0.5 m thick. Its lower section is layered with crushed stones, pebbles, and air ducts and is generally designed with a linear velocity of 3 to 5 mm/s. This method is easy to maintain, relatively inexpensive to operate, and not dependent on the actual shape of the land. Nevertheless, it requires a large land area, and it is necessary not only to weed but also to plow or replace the soil to prevent air conduits from forming.[19]

The highest level of odor control is provided by running the exhaust air through thermal afterburners. This process, also known as thermal regenerative oxidation, exposes the malodorous airstream to temperatures ranging from 1400 to 1800°F with a residence time of 1 s or more at that temperature. This is a costly but very effective method of controlling odors.

9.4 MARKETS FOR COMPOSTED SLUDGE

The primary factors that impact compost marketing are the type of amendment used, the final compost screening, and the level of effort associated with the marketing program. Every potential compost market requires product quality standards involving such characteristics as the absence of pathogens, particle size and texture, moisture content, biological stability determined by oxygen demand, odor, weed seeds, pH, and concentrations of nutrients, salts, and metals. High-quality uses include retail sales of bagged compost to the general public, sales to the nursery industry for potting soils, and sales to landscape contractors and public agencies for use as top-dressing. Lower-quality compost can be used as landfill cover or to reclaim disturbed land.

Experience has shown that a facility that produces a high-quality compost can market the material, but it doesn't happen automatically. Without a successful marketing program, the compost will end up in stockpiles. Compost doesn't sell itself until the end-product users are educated to understand its benefits. Many treatment plants opt not to develop a marketing program and instead hire a private broker to dispose of their finished compost. Private brokers already understand the benefits of using compost and are often associated with large-volume soil users. They can develop broad markets for compost, ranging from golf courses to nurseries.

A first step in developing a marketing program is to use compost for public works projects with high visibility. Typical demonstration projects include public parks and gardens, median strips, traffic rotaries, and the grounds of public buildings. Successful projects should have signs that identify the site as a compost demonstration project.

Successful compost marketers understand the proper use of compost and convey this information to end-product users. They maintain communication with users to help overcome any difficulties and to publicize their successes. Many POTWs conduct research projects to establish the benefits and optimal application rates for their compost. Research projects increase the credibility of the composting program while yielding useful information.

Once interest in the material has been established, local markets must be identified. Public agencies use compost for park maintenance, recreational areas, cemeteries, road embankments, and building grounds. Workers apply compost as mulch around trees and shrubs or as topdressing to lawns. Large one-time uses of compost include the construction of baseball fields and cemeteries and the installation of roadside curbing. Another large one-time use is land reclamation, including exhausted gravel pits, sand pits, strip mines, urban lots, construction sites, and other disturbed areas. By far the largest bulk use of compost by the public sector is the operation and closure of landfills. Compost can be used as grading material prior to capping and to develop vegetative support layers on capped landfills.

General and landscape contractors are also a major market for compost. Once the product is publicized, selling or giving compost to landscape contractors can result in lines of trucks outside the composting facilities. Where markets have been established, loam production can represent a significant compost market. By mixing compost with sandy soil in a volumetric compost-to-soil ratio of from 1:2 to 1:5, compost can be transformed into loam.

Homeowners can represent a significant market, where recipients typically pick up their compost in containers ranging from 5-gal buckets to pickup trucks. The drawback to this market is the amount of time that the operators have to spend helping customers load containers. Educating small users on the proper use of the product is also time-consuming.

The most popular method of compost distribution is pickup at the composting plants. Most facilities load bulk-user vehicles, while small-quantity users must load their own. Some facilities deliver compost locally for a modest fee.

Table 9-9 gives a sampling of a few composting facilities around the country and the compost end uses. All of the compost produced by these facilities is considered to be a "good" product, meeting local regulatory requirements and having a demonstrated usefulness as a soil conditioner.

Table 9-10 presents typical selling prices of finished compost. The price for finished compost varies considerably, from $1.35 to $16/yd^3. Many facilities give away their compost for free to anyone, homeowner or contractor, who will come and pick up the material.

9.4.1 Akron, Ohio

The Akron, OH compost facility began making compost in 1988. The first year they produced 40,000 yd^3 of material and sold 17,000 yd. That year they established an advisory council of technicians and potential compost consumers to discuss compost marketing. The potential consumers discussed their needs and pricing structures and outlined areas where new research was needed. The managers of the facility had the opportunity to allay fears about compost odor and safety and to acquaint users with application rates and the daily, weekly, and monthly monitoring programs that maintain compost quality. One result of these meetings was allocation of about $30,000 in compost research funds, mostly to Ohio State University. Akron managers concluded that there were two distinct markets: the commercial bulk customer and the residential home market. An advertising agency was used to find names for the products and came up with Soil Magic™ for the home market and TechnaGro™ for commercial outlets.

The second year the facility produced 35,000 yd^3 of compost and sold 37,000 yd, bringing in $162,000. About 50% of the compost went to topsoil blenders, nurseries accounted for about 35% of the sales, and 15% went to landscapers and homeowners. The prices ranged from $16.00/yd for from 1 to 60 yd on down to $4.50/yd for from 181 to 500 yd and $4.00 for anything over 500 yd. There was a big gap in the volume of compost used by commercial and residential users: one-half went to three topsoil

Table 9-9 Compost Markets in the United States

Location	Volume (cubic yards)	Product name	Method of distribution	Typical users
Akron, OH	32,000	Soil Magic	Direct sales	50% —Topsoil blenders
		TechnaGro		35% —Nurseries 15% —Landscapers homes
Clayton County, GA	5,000	None	North American soils	75% —Landscapers 10% —Nursery
Hampton Roads (VA)	11,000	Nutri-Green	Direct sales	40% —Landscapers 30% —Municipal 20% —General public
East Bay MUD, WA	25,000	CompGro	Ten Brokers	Landscapers
Los Angeles County Sanitation District, CA	250,000–300,000	Nitrohumus Topper, Gromulch Amend, and others	Kellogg Supply Company	General public and nurseries
Nashville, TN	9,000	Nitro Humus	Soil Products Inc.	90% —Landscapers and nurseries
Scranton, PA	15,000	Pay Dirt	Direct sales and single broker	25% —Landscapers 25% —Municipal give-away
Washington Suburban Sanitary Commission, MD	40,000	ComPRO	Maryland Environmental Service	25% —Landscapers 40% —General public 25% —Topsoil dealers

Sources: Donovan[22] and Logsdon[20]

Table 9-10 Compost Selling Prices

Location	Price ($/yd^3)	Per volume sold
1. Akron, OH	16	<60 yd^3
	5	60–180 yd^3
	4	>180 yd^3
2. Clayton County, GA	6	
3. Metropolitan Denver	8	Delivered up to 20 miles
Sewage Disposal, CO	16	Delivered up to 100 miles
4. Hampton Roads	12	<10 yd^3
Sanitation District, VA	8	10–500 yd^3
	6	>500 yd^3
	1.35	1 ft^{3}bag
5. Nashville, TN	5	
6. Scranton, PA	3	
7. Washington County Sanitary Commission, MD	7.50	
9. Bangor, ME	5	
10. Merrimack, NH	2	

Sources: Donovan[22] and Lang and Jager.[17]

dealers, while 300 homeowners bought 2% of the product. To provide more market stability, a green and black bag was designed to help increase sales to homeowners. These 40-lb bags were sold for $1.00. The demand has been steadily increasing from 3,000 in 1988 to 13,000 in 1989 and 25,000 in 1990. To further diversify the product an extra-fine material for lawn application was added in 1990, selling at $1.50 per bag.[20]

The Ohio Department of Natural Resources provided a large demonstration project for Akron when they took 25,000 yd^3 of compost in 1989 to rehabilitate strip-mined land in the southern part of the state. The strip-mined sites had been left with the subsoil on the surface. Without the organic matter and micronutrients present in topsoil, plants don't grow and the land doesn't revegetate. Without vegetation the soil erodes, causing surface and groundwater degradation.

When sludge is used to stabilize and revegetate land, enormous amounts — up to 112 t/ha or more — are applied on a one-time basis. This large amount is necessary to ensure that sufficient organic matter and nutrients are introduced to the soil to support vegetation until a self-sustaining ecosystem is established. Akron's sludge compost returned organic matter to the soil and provided a base on which a permanent vegetative cover could reestablish itself. While Akron got no money for this transaction, it had access to use the site for sludge research and reduced its stockpile of compost to zero.

Once the strip-mined site is permitted by the state, the land is limed to neutralize the soil pH. Below a pH of 6 the metals in the soil are free to be taken up by plants and to leach into the groundwater. Lime application raises the pH to 7, binding the metals so they are unavailable for plant uptake. Several inches of sludge compost are applied to the neutralized soil, restoring organic matter and building topsoil from subsoil. The site is then sown with perennial grasses and legumes. The soil and groundwater at the compost application sites are monitored to study the environmental effects of the sludge application, and if the pH of the soil begins to drop the land is relimed.

Hills that were gullied, eroded, and barren are now healthy, green, and permanently revegetated. Wildlife has returned, the pollution of the groundwater and surface water has ceased, and land that once looked like waste heaps has become verdant hills. As goes the demonstration project, so goes the marketing program. Akron's compost is black gold.

REFERENCES

1. Goldstein, N. "Cocomposting Sludge and Yard Waste," *Biocycle* (January 1991), pp. 31–34.
2. Burnett, C.H. "Small Cities + Warm Climates = Windrow Composting," Presented at the Water Environment Federation 65th Annual Conference and Exposition, New Orleans, LA (September 20–24, 1992).
3. Bennett, L., E. Epstein, and W. Porter. "Sludge Composting Comes to Idaho," *Biocycle* (February 1991), pp. 54–56.
4. Harris, G.D., W.L. Platt, and B.C. Price. "Vermicomposting in a Rural Community," *Biocycle* (January 1990), pp. 48–51.

5. Loehr, R.C., J.H. Martin, and E.F. Neuhauser. "Liquid Sludge Stabilization Using Vermistabilization," *J. Water Pollut. Control Fed.* 57(7):817–826 (1985).
6. Hay, J.C. and R.D. Kuchenrither. "Fundamentals and Application of Windrow Composting," *J. Environ. Eng.* 116(4):746–763 (1990).
7. Kuchenrither, R.D., W.J. Martin, D.G. Smith, and D.W. Williams. "Design and Operation of an Aerated Windrow Composting Facility," *J. Water Pollut. Control Fed.* 57(3):213–219 (1985).
8. Johnston, J.R., J.F. Donovan, and A.B. Pincince. "Lessons on In-Vessel Composting," *Water Environ. Technol.* (May 1990), pp. 56–63.
9. Walker, J., N. Goldstein, and B. Chen. "Evaluating the In-Vessel Composting Option, Part I," *Biocycle* (April 1986), pp. 22–27.
10. Walker, J., N. Goldstein, and B. Chen. "Evaluating the In-Vessel Composting Option, Part II," *Biocycle* (May-June 1986), pp. 34–39.
11. Rutland, L. and R.M. Ryan. "In-Vessel for Sludge and Yard Waste," *Biocycle* (September 1991), pp. 56–57.
12. Newman, M.L., J. Brandon, and C. Gregory. "Closing the Water and Sludge Loop," *Biocycle* (February 1989).
13. Logsdon, G. "Sludge Composting Under Way in Hamilton!, Ohio," *Biocycle* (August 1989), pp. 39–41.
14. Wilber, C. and C. Murray. "Odor Source Evaluation," *Biocycle* (March 1990), pp. 68–72.
15. Murray, C.M. and J.L. Thompson. "Odor Control at Composting Facilities: Strategies for Aerated Pile Systems," *Biocycle* (July 1986), pp. 21–26.
16. Hay, J.C. et al. "Alternative Bulking Agents for Sludge Composting," *Biocycle* (October 1988), pp. 48–51.
17. Lang, M.E. and R.A. Jager. "Compost Marketing in New England," *Biocycle* (November 1990), pp. 68–71.
18. Jager, R.A. and M.E. Lang. "Evaluation of Odor Control Technologies for Municipal Sludge Composting," Presented at the Water Environment Federation 65th Annual Conference and Exposition, New Orleans, LA (September 20–24, 1992).
19. Kaneko, S. and K. Yamanaka. "Environmental Measures in Sewage Works," *Water Environ. & Technol.* (May 1991), pp. 59–62.
20. Logsdon, G. "Selling Sludge Compost," *Biocycle* (May 1990), pp. 75–77.
21. Goldstein, N. and D. Riggle. "Sludge Composting Sets Healthy Pace," *Biocycle* (December 1991), pp. 30–37.
22. Donovan, J.F. "Markets for Sludge Compost," *Biocycle* (February 1990), pp. 44–45.
23. Iacoboni, M.D., "Windrow and static pile composting of municipal sludges." Report for the U.S. EPA, Sanitation District of Los Angeles County, Los Angeles, CA (1983).

10 SLUDGE PELLETIZING

10.1 INTRODUCTION

Dewatered sludge cake can be processed further by pelletization. The pelletizing process heat dries the sludge to a water content of about 5%, reducing the volume of the dewatered cake by a factor of roughly 15. The resulting sludge product can be bagged and distributed across the country as a soil amendment or fertilizer.

Heat-dried sludge, either in pellets, powders, or flakes, has been marketed nationally as a fertilizer for nearly 70 years. Milorganite, the sludge pellets produced and packaged by the Milwaukee Metropolitan Sewerage District, has been sold since 1923. In 1989, Milorganite was available in all 50 states, the Caribbean, and Canada. It is sold in local lawn and garden stores in bulk or in bags, the majority of the sales being in bulk. About one third of the product is sold in 50-lb bags to commercial turf growers, one third is sent to Florida and the southeastern U.S. for direct application or blending, and the remainder is sold in 40-lb bags for home use. With the advent of the new sludge regulations and the increasing amount of pelletized sludge produced across the country, Milwaukee's marketing program is likely to change in the coming years.

There is considerably higher public enthusiasm for sludge pellets than for any other sludge product. Pelletized sludge product is virtually odor-free and is interchangeable with commercial fertilizers: it looks, feels, and smells like any other organic fertilizer. The inoffensive nature of the product and the broad acceptance and widespread use of Milorganite have made pelletized sludge a familiar and trusted product.

In the sludge pelletizing process, dewatered sludge cake is heat dried to produce pellets, powders, or flakes. The solids content of pelletized sludge ranges from about 90 to 98%, so there is less sludge product to be marketed. Pelletizing produces the lowest volume of sludge product of any of the processing technologies that allow sludge to be beneficially reused. Pellets are a highly marketable commodity and have a low enough water content to be easily transported. The pelletizing process requires a relatively small processing site and offers good potential for odor control. Heat drying technology solves most problems associated with the beneficial reuse of sludge: there is a minimal volume of sludge product, the processing odors are easily contained for treatment, there is little land area required for a processing site, and the product is readily accepted by the public.

For pelletized sludge to meet class A standards, the moisture content of the pellets must be 10% or lower. In addition, either the temperature of the gas in contact with the sludge must exceed 80°C or the wet bulb temperature of the gas in contact with the sludge as the sludge leaves the dryer must exceed 80°C. For a class A sludge product, in addition to treatment, either the density of fecal coliforms in the sludge must be less than 1000 MPU/g total dry solids or the density of *Salmonella* sp. bacteria in the sludge must be less than 3 MPU/4 g total dry solids at the time the pellets are sold or used. The vector attraction reduction requirement is met by pellets of at least 90% total dry solids if the feedstock includes unstabilized primary solids. If the pellets do not contain unstabilized primary solids, a total dry solids content of 75% is sufficient to meet the vector attraction reduction requirements. In general, high levels of heavy metals are the only reason that a well-designed pelletizing process will produce class B sludge.

10.2 PELLETIZING PROCESS

The objective of all heat drying processes is the evaporation of water from the sludge, but the equipment, materials handling, energy requirements, and products differ significantly with each pelletizing system. Heat drying can be carried out with either indirect or direct heat.[1] Figure 10.1 shows a schematic of the materials input and output from an indirect dryer and a direct dryer.

Pelletizing systems are composed of subsystems which interact to create sludge pellets. These subsystems can be grouped by function as follows:[1]

- Sludge cake feed
- Heat source
- Sludge dryer
- Product post-processing
- Airstream treatment

The sludge cake feed systems typically include sludge cake storage hoppers, enclosed screw conveyors, and a pug mill. The dewatered cake is transferred by screw conveyor from the storage hopper to the pug mill, along with the dried material from the recycle bin. In the pug mill, the recycled dried material and the sludge cake are thoroughly mixed until they become a homogeneous material that is considerably drier than the initial sludge cake. The blended material from the pug mill is about 70% solids and is now ready for heat drying.

In direct heat drying, a heated airstream in contact with the wet sludge evaporates moisture and dries the sludge. The airstream can be heated in a variety of ways, including waste heat, solid material, and natural gas. The sludge cake and recycled product mixture is usually transferred to the rotary drum dryer via screw conveyor, which allows optimal control of the rate that sludge cake is delivered to the drum. The dryer drum is typically a single- or triple-bypass cylindrical vessel that rotates at approximately 20 rpm. There the sludge is contacted by a stream of hot air at about 700°F. The sludge moves through the dryer by means of the airflow and drum

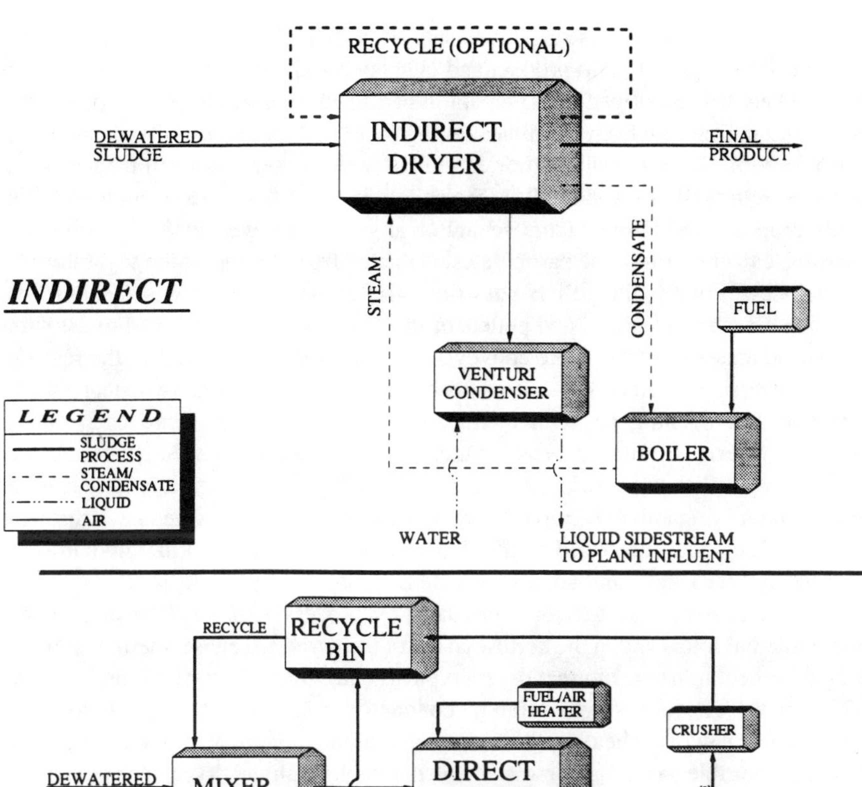

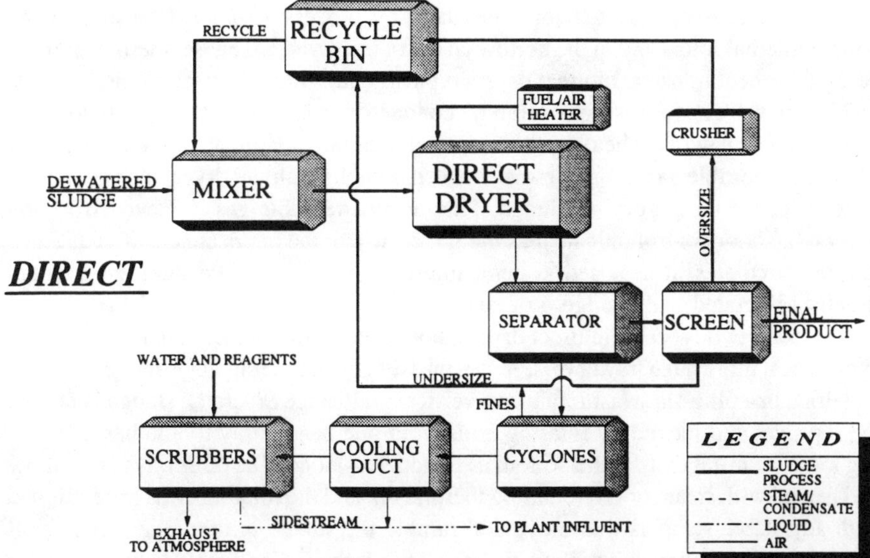

Figure 10-1 Sluge pelletizing schematics of indirect and direct dryers. (From Psaris, P. J., Roll, R. R., and Schatz, A., "Pilot Testing: Lessons Learned Evaluating Heat Drying for Residuals Management," WEF 65th Annual Conference, 1992. With permission.)

rotation. The hot airstream transports the sludge through the center passage of the rotating dryer, which consists of one cylinder or three concentric horizontal cylinders. For triple-pass dryers, the pellets then reverse direction to be transported through the middle passage and reverse direction again to be transported through the outside passage.

The sludge product at this stage consists of small, hard granules of varying sizes which must be separated into product and oversized and undersized material. To do so, the pellets are discharged to the separator can and the screening device for further classification. The airstream continues through the separator can, cyclone, and odor control system. The airstream and dried pellets enter the separator can tangentially, where the velocity is slowed to allow the pellets to separate from the airstream. The pellets drop to the bottom of the separator and are conveyed to the screen. The airstream, carrying some fine particulates, is ducted from the top of the separators to the cyclones.[2] Finally, the pellets pass through stainless steel mesh screens in the screening device to be sized, and pellets of the proper size are conveyed to the silo. Undersized material, or fines, are conveyed from the screening device to the recycle bin. The oversized material is sent to the crusher, which contains two steel rollers with a narrow opening between them. As the rollers rotate toward each other, oversized material is crushed between them before being transferred to the recycle bin. The recycle bin stores the dried material from the crusher until it is conveyed to the pug mill mixing unit to be mixed with dewatered sludge and cycled back through the driers. The pellets produced by this process are roughly 1.5 to 4 mm in diameter, about 95 to 99% solids, and have a bulk density of about 35 to 45 lb/ft^3.

An indirect drying system provides heat by process steam condensing in the hollow internal rotors and in the hollow jacket of the dryer. Likewise, thermal oil may be used for heat transfer. Indirect dryers typically have high thermal efficiencies (up to 75 to 80%). The heat is transferred by conductivity, and since the sludge does not come in contact with the heating medium it remains uncontaminated. A small amount of noncondensable gas, mostly moist air, is drawn through the dryer to carry out the water vapor from the drying sludge. This noncondensable gas is vented from the dryer to an odor control unit such as the boiler, where the odors are destroyed. Since the gas circulates in a closed system, indirect dryers avoid the unpleasant odors usually associated with sludge processing.

There are two types of indirect dryers: horizontal screw-mixer "en masse" dryers and vertical multistage tray dryers. In "en masse" drying, the sludge is moved, mixed, and dried in a plug-flow fashion. In the vertical multistage dryer, the sludge is fed via the top inlet and moved by rotating arms from one heated tray to another in a zig-zag motion until it exits at the bottom as pelletized sludge. The drier trays are hollow and heated with steam or recirculating thermal oil, and the rotating arms are equipped with adjustable scrapers that move and tumble the sludge in thin layers and small windrows. The drying and pelletizing process starts with fine particles which gradually increase in size, layer by layer, growing larger as they dry from the center to the outside. In general, the pellets produced by indirect dryers tend to be fluffy, fibrous, and nonuniform when compared to the pellets produced by direct dryers.

10.3 PELLETIZING COSTS

The pelletizing process requires a substantial amount of equipment and is a capital-intensive method of processing sludge. The costs of sludge pelletizing at the Clayton County, GA facility are shown in Table 10-1. The construction costs for the

Table 10-1 Costs of Pelletizing Sludge at Clayton County (GA) Water Authority

	Annual costs	$/dry ton processed
Operating expenses		
Direct labor and benefits	$172,230	$59.55
Electricity	73,525	25.42
Other utilities	81,780	25.42
Wood chips	44,578	15.41
Dewatering chemicals	176,906	61.17
Supplies	64,812	22.41
Outside services	111,998	38.73
Maintenance and repairs	41,206	14.25
Miscellaneous and overhead	53,224	18.40
Total operating expenses	820,259	283.62
Product credit from sales	209,592	72.47
Net operating expenses	610,667	211.16
Interest on $651,500 of facility	49,991	17.29
Depreciation expense	130,427	45.10
Net costs	$791,427	$273.66/dry ton

Source: Newman, Brandon, Gregory.[3]
Note: Based on 2892 dry tons processed, FY 1988, EPA grants = 83% of construction.

facility, including the pelletizer, incinerator, wood chip handling equipment, and buildings, totalled $3,912,829. EPA grants paid for $3,261,329 of the construction costs, so nearly 85% of the plant is owned free and clear with no interest payments. The interest payments in Table 1 represent interest on the $651,500 not covered by the various EPA grants. To provide a clearer picture of the true costs, the entire facility is depreciated on a straight-line method with a 30-year life. However, they do not include interest on the EPA's $3.26 million.[3]

The costs of pelletizing sludge in Burlington, VT are given in Table 10-2. The state subsidy for the construction costs of this proposed facility totals 40%, giving a higher cost per dry ton than in Clayton County; however, this is still well below the real costs. Since this cost analysis doesn't include dewatering costs, the actual costs of pelletizing would be closer to $340/dry ton.

Finally, the sludge pelletizing facility in Boston is operated by the New England Fertilizer Company (NEFCO), a joint venture between Baltimore's Enviro-Gro Technologies and three other engineering, construction, and large project financing companies. NEFCO will pelletize Boston sludge for 4 years at the contractual prices given in Table 10-3. Here we see the costs of pelletizing sludge without state and federal subsidies and with enormous economies of scale.

10.3.1 Pelletizing in Clayton County, Georgia

The Clayton County Water Authority produces 55 dry tons of sludge pellets per week.[3] The pelletizing process train was installed by Enviro-Gro Technologies in 1979. To produce the BB-sized pellets, the Authority first dewaters the waste-activated sludge from an incoming 3.5% solids to an outgoing 18% solids on two belt filter presses. Dewatered sludge is further dried by mixing with previously dried

Table 10-2 Costs of Pelletizing Sludge in Burlington, Vermont; No Dewatering Costs

	Annual total	$/dry ton processed
Operating expenses		
Direct labor and benefits	$110,000	$40.56
Power	336,000	123.89
Fuel	4,000	1.48
Water and wastewater	12,000	4.43
Equipment replacement	28,000	10.32
Maintenance	95,000	35.03
Miscellaneous and overhead	59,000	21.76
Total operating expenses	644,000	237.46
Product credit from sales	162,720	60.00
Net operating expenses	481,280	177.46
Annualized construction costs	331,655	122.29
Net costs	$791,427	$299.75/dry ton

Source: Adapted from Dubois and King.[5]
Note: Based on 2712 dry tons processed, FY 1993, state grant = 40% of construction.

Table 10-3 Costs of Pelletizing Sludge in Boston, MA; No Dewatering Costs

Quantities of sludge processed	Cost/dry ton
First 900 dry tons/month	$358.06
Next 600 dry tons/month	294.57
Next 600 dry tons/month	269.37
All over 2100 tons/month	255.17

Source: MWRA.[2]
Note: 1992–1995, negligible grants or subsidies.

sludge pellets, fines, and regrind at a 3:1 ratio of recycled to new to reach a solids content of 70% before the material enters the dryer. The material is mixed in a double-rotor pug-mill mixer before it is introduced to the dryer. The dryer is a triple-pass rotary kiln. The sludge/pellet mix is drawn through the kiln by a fan, along with hot air from the wood chip incinerator at 17,000 ft^3/min. The sludge mixture dries into pellets as it rolls through the kiln.

The system is fueled by wood chips burned in a pulse hearth incinerator system, which can reburn combustion air. This feature was originally designed to allow the process to be fueled by sludge pellets, which were not seen as a particularly valuable commodity in 1979. Subsequent research opened the agricultural market for sludge pellets, precluding their use as a fuel.

The sludge pellets are sized for the agricultural market. About 85% of the pellets come out of the drier within size specification limits. They are either shipped as fertilizer or recycled to dry the incoming sludge. The remaining 15% oversized fraction is crushed and rescreened. Less than 1% of the processed sludge is reduced to fines and dust, which go back to the head of the process train. About 15% of the incoming sludge is lost during the pelletization process.

The sludge product, Agri-Plus 650®, is a registered fertilizer with a guaranteed analysis of 6-5-0 nitrogen-phosphorus-potassium content. The Authority sells all output to Lykes Agri-Sales at a contract price of $82/ton of pellets, which translates to $72.47/dry ton of sludge. Lykes markets the pellets to the Florida orange grove trade directly and also uses the pellets as a base material for a stronger, more complete fertilizer containing potash. Both products are class A fertilizers that can be used on first-chain food crops.

10.4 MARKETS FOR PELLETIZED SLUDGE

Sludge pellets can be marketed aggressively, although the income from product sales may not do much more than cover the costs of the marketing program. Sludge pellets can be used in the following markets:

1. Turf fertilization (production, commercial, residential)
2. Nursery (landscape plant stock) fertilization
3. Ingredient in growth medium products (e.g., potting soil)
4. Silviculture (forest fertilization)
5. Soil conditioner
6. Bulking agent or filler in dry blended fertilizers
7. Fertilizer for export

These uses have several different levels of market return. Using sludge pellets for turf fertilization should offer a higher return than silviculture sales. Lower return markets are typically larger and far more reliable than the markets that maximize revenue.

Table 10-4 provides an overview of the prices that sludge pellets are sold for in the U.S. Over the next few years, as more sludge pelletizing facilities come on line, the price for the pellets is likely to drop as the market becomes saturated.

In order to market pellets successfully, the following major criteria[4] should be met:

1. Moisture content. The pellets should not have moisture in excess of 5 to 6%. Higher moisture causes mold and microorganism growth, overheating, and possibly fires.
2. Nutrient content. Pellet fertilizer values depend primarily on nitrogen and phosphorus content, nutritional elements which are always present in sludge. Total organic and inorganic nitrogen content may be in the 1.5 to 6% range by dry weight. The total organic matter of the pellet may be in the 35 to 65% range.
3. Particle size distribution. From a marketing and distribution standpoint, the pellets should be homogeneous in size with a narrow particle size distribution in the 1.0 to 3.0 mm range.
4. Durability. The pellets should not break and produce dust when handled. A standard product durability index (PDI) should generally be between 92 and 98 units.
5. Microbiological and bacteriological testing. The pellets should comply with the applicable standards concerning pathogenic and indicator organisms. Pellet-size and low-moisture-content uniformity are important factors in the microbiological "cleanliness" of the pellet. Elevated moisture inside the chunky material contributes to microorganism survival and growth.

Table 10-4 Pelletized Sludge Products in the United States

Products available	Price per ton F.O.B. Florida[a]	Price per unit of nitrogen[a]
Municipal sludge (not enriched)		
Milorganite®, bagged (Milwaukee, WI)	$145	$24
Houactinite®, flakes (Houston, TX)	120	22
Agri-Plus 650®, granular (Clayton County, GA)	85	14
Enviro-Gro®, granular (private)	80	20
Industrial sludge or enriched sludge		
Hynite®, granular (leather sludge, 11% N)	186	17
Organiform®, granular (Hynite with urea added, 24% N)	313	13
Charmin®, granular (paper mill sludge, urea enriched, 8% N)	150	19
Chemical fertilizers		
Sulfur coat urea	335	9
Granular ureaform	605	16
Ureaform	510	13

Sources: Girovich[4] and Newman, Brandon, and Gregory.[3]
[a] Prices based on June 1989 figures.

6. Heavy metals and hazardous organics. Sludge metal content in pellets is the most important factor in marketing and distribution. With the possible exception of mercury, the metals present in the sludge feed will generally remain in the pellets.

There are a number of different levels of pellet utilization.[2] Pellets can be used as a fertilizer, with their price based on the market value of nitrogen, potassium, and phosphorus. The baseline for quality heat-dried sludge is currently established by Milorganite. To be an effective fertilizer pellet, the sludge product may have to be blended before marketing. The fertilizer value of sludge pellets can be enhanced in a number of ways: pellets can be mixed with standard fertilizer ingredients or soil conditioning materials or coated with other fertilizer ingredients.

If the sludge pellets have very low plant nutrient analyses, they may be useful as an agricultural-grade fertilizer filler. Granular limestone is typically used for this application, at a cost to the fertilizer manufacturer of between $15.00 and $25.00/ton delivered. The limestone has limited plant nutrient value, since the size of the limestone granule makes it an inefficient source of calcium and an ineffective pH buffer for the soil. Sludge pellets would be a preferable filler for fertilizer manufacturers, since they contain some percentage of water-insoluble nitrogen. Using pellets as filler, the manufacturer can reduce the outlay for synthetic chemical fertilizer while adding an organic component to the fertilizer.

Sludge pellets can be exploited for their organic nutrient base rather than their fertilizer value. This market includes plant growth media, mulches, composts, and other landscape products. The sludge pellets can be mixed with routinely used low-nutrient commodities. The "soilless" potting medium for seedlings, for example, is usually a mixture of equal parts sand, peat moss, and vermiculite, products which offer little plant nutrient value. Pellets are used primarily for their organic contribution; the addition of sludge pellets to peat moss, composted manure and sludge, and organic bark mulches would offer some fertilizer value while enhancing the soil properties of the mulch.

The worst-case scenario for sludge pellets would be the local distribution of free sludge pellets and donations of pellets to local government groups. This could occur for two reasons: the fertilizer value of the sludge is so low that the pellets can be used only for their organic properties, or the physical characteristics of the pellets are unacceptable to the fertilizer industry. Turf amendment in public parks, establishment of planting in local public works projects, reclamation of urban land and abandoned lots, rejuvenation of pastureland, and the fertilization of public forests are all plausible worst-case methods of utilizing large quantities of sludge pellets, offering little or no return on product sales.

REFERENCES

1. Psaris, P.J., R.R. Roll, and A. Schatz. "Pilot Testing: Lessons Learned Evaluating Heat Drying for Residuals Management," Water Environment Federation 65th Annual Conference and Exposition, New Orleans, LA (September 20–24, 1992).
2. "Interim Sludge Processing and Disposal Project, Final Facilities Plan, Environmental Impact Report, Volume II," Prepared by Tighe and Bond, Inc. for the Massachusetts Water Resources Authority (April 1989).
3. Newman, M.L., J. Brandon, and C. Gregory. "Closing the Water and Sludge Loop," *Biocycle* (February 1989).
4. Girovich, M.J. "Simultaneous Sludge Drying and Pelletizing," *Water Eng. Manage. Mag.* (March 1990).
5. "Sludge Management Plan Draft Report," Prepared by Dubois and King for the Chittenden Regional Solid Waste Management District, Vermont (1991).

11 INNOVATIVE TECHNOLOGIES FOR SLUDGE REUSE

11.1 INTRODUCTION

Innovative technologies for sludge reuse stretch the boundaries of what is traditionally considered to be beneficial reuse. If a process requires a reasonable input of energy and creates a usable sludge product, than it can be considered by some definitions to be a beneficial reuse.

A variety of innovative sludge processing technologies have been developed in the last decade, some of which have not yet been proven in a full-scale installation. It is quite difficult to bring laboratory technology to a full-scale municipal treatment plant. The costs of constructing new sludge processing facilities are enormous, and metropolises with daily sludge disposal problems are understandably risk averse. The sludge processing technologies described in this chapter have had extensive pilot testing in the U.S. or have been successfully implemented abroad, but most of them have not been proved with a full-scale facility in this country.

11.2 SLUDGE TO OIL — STORS

In the 1930s, German researchers discovered that heating biomass and treating it with alkali produced a charred material they called artificial coal. In the early 1980s scientists from the Columbus, OH-based Battelle research facility at the Battelle Pacific Northwest Laboratory in Richland, WA focused on this old German work using sludge as the feedstock. Funding was provided by the U.S. EPA, the U.S. Department of Energy, and a Japanese company. The result was the STORS process, an acronym for sludge-to-oil reactor system.[1]

11.2.1 Process

The STORS process, shown in Figure 11-1, takes sludge at 4% solids, centrifuges it to 20% solids, and then adds about 5% sodium carbonate alkali (washing soda) as a catalyst to assist molecular rearrangement.[1,2] The organic mixture is then pumped into a thermochemical reactor, a sort of pressure cooker. STORS is called a "high-pressure poop pump" by its inventors, since the sludge is heated to about 300°C and

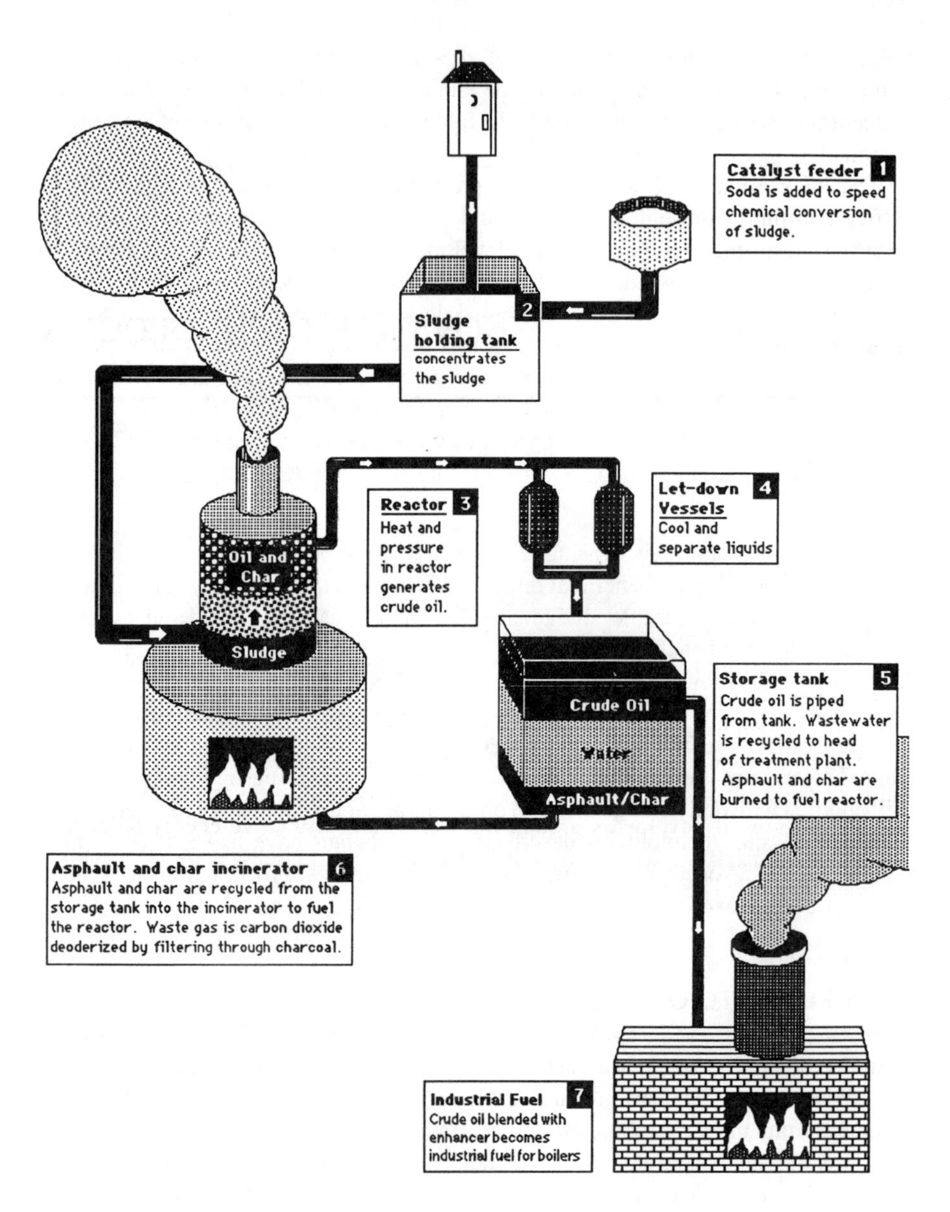

Figure 11-1 Sludge-to-oil reactor system: STORS. (From Case, F., "Waste Not," *The Seattle Times,* Oct. 3, 1988. With permission.)

held for less than 1 h under its own steam pressure — about 2000 psi. This sacrifices the carbon to remove the oxygen, rearranging the hydrogen and carbon molecules to create a hydrocarbon. When cooking is complete, the product is decanted into let-down vessels that cool the sludge and separate it into four components: wastewater, a sludgy char made up of organic matter used as a fuel to heat the reactor, an off-gas that is 90% carbon dioxide, and, floating on top of the water, an odoriferous black oil that looks like something you'd drain out of your car after 3000 miles. The heating value of the oil is 80 to 90% that of diesel fuel, and it is relatively clear of contaminants. Most of the metals in the sludge are transferred to the char; metals concentrations in STORS oil are up to 27-fold lower than the starting sludge concentrations.[3]

The oil itself has a few unfortunate properties: it is a thick, black goo that is difficult to pump through feed lines, and it smells an awful lot like sludge. To improve these qualities, an additive developed by American Fuel and Power in Panama City, FL is now added to the oil to cut the viscosity and sweeten the smell. The improved oil looks and pours like no. 4 diesel fuel. If deodorized, it can be used for ship or industrial boilers or even distilled and upgraded for use in automobiles.[4]

American Fuel provided the additive that upgraded the oil to a usable commodity and the money to complete the demonstration project. In exchange, they got the license for the STORS process. The worldwide marketing rights for the STORS process are now controlled by a spin-off company: Innotek Corporation of Little Rock, AR.[5]

One of the most appealing attributes of the STORS process is that it effects an enormous reduction in sludge volume. If the process begins with sludge that is 20% solids, 500 tons of sludge are transformed into 30 tons of ash, considerably decreasing the volume of residuals that must be disposed of.[6] The products of the STORS process are shown in Figure 11-2. The liquor is returned to the head of the treatment plant, the char is used to power the process, and the oil is burned for fuel, leaving only the ash to be landfilled.

The energy required to fuel the extraction process is only one third of the energy value of the products, a factor that sets STORS apart from the approximately 200 other sludge-to-oil developers around the world. This attractive energy balance is attributable partially to the use of a wet feedstock and partially to the fact that the products separate spontaneously.[7]

11.2.2 Costs

After a decade of laboratory development and a successful pilot plant, the STORS technology has yet to be implemented. Partially because the municipal application of the process is still theoretical, the costs of constructing and operating a STORS facility have been carefully projected. Shown in Table 11-1 are three scenarios: the costs for a town of 10,000, a city of 100,000, and a metropolis of 1 million in 1990 U.S. dollars.

11.3 OIL FROM SLUDGE

The primary differences between the oil from sludge (OFS) process and the STORS process is that OFS uses dried sludge pellets as the feedstock while STORS uses dewatered sludge, and OFS is a low temperature-atmospheric pressure process while STORS is a high pressure-high temperature process. Like STORS, the OFS technology thermally converts the volatile organic matter in sludge to a liquid fuel predominantly consisting of straight-chain alkanes and alkenes, much like the principal

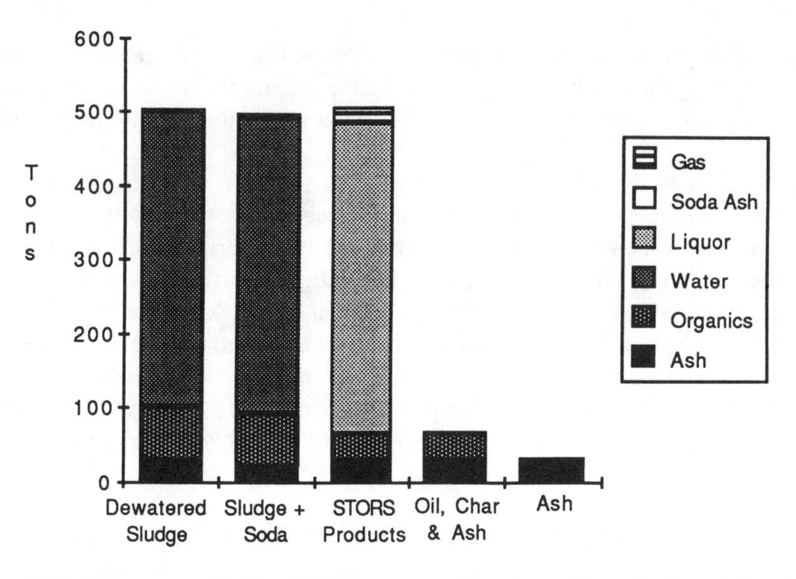

Figure 11-2 Volume reduction by STORS process. (Courtesy of Innotek Corporation, Little Rock, AR)

Table 11-1 Costing STORS for Three Facility Sizes

	Town of 10,000		City of 100,000		Metropolis of 1 million	
Item	**% of capital cost**	**Estimated cost**	**% of capital cost**	**Estimated cost**	**% of capital costs**	**Estimated cost**
Capital cost for STORS						
Engineering design	15	$100,000	13	$200,000	8	$500,000
Materials	68	$470,000	66	$1,031,500	69	$4,466,500
Construction	15	$100,000	19	$290,000	17	$1,150,000
Start-up	1.02	$7,000	0.64	$10,000	3	$200,000
Indirect capital cost	1.46	$10,000	1.48	$23,000	3	$170,000
Total capital costs	100	$687,000	100	$1,554,500	100	$6,116,500
	% of operating cost	**Estimated annual cost**	**% of operating cost**	**Estimated annual cost**	**% of operating costs**	**Estimated annual cost**
Operating costs:						
Debt service at 8%, 20 years	39.9	$69,972	41	$158,329	35	$660,664
Maintenance, labor, and supplies	14.2	$25,000	6	$25,000	5	$100,000
Operator labor	22.8	$40,000	21	$80,000	13	$250,000
Fuel and soda ash	8.5	$15,000	26	$100,000	40	$750,000
Miscellaneous	14.5	$25,500	6	$25,600	5	$106,400
Annual operating costs	100	$175,472	100	$389,929	100	$1,867,064
Annual sales revenue	—	$0	—	$55,672	—	$556,719
Sludge disposal costs ($/dry ton)	—	$601	—	$114	—	$45

Source: U.S. EPA.[2]

ingredients of diesel fuel. The by-products of the process are char, noncondensible gas, and reaction water, which are combusted on-site to supply the energy requirements for drying the sludge and heating the reactor. The oil can be used as a diesel fuel equivalent.

About the same time as Battelle research engineers developed the STORS process, German researchers at Tübingen University made significant advances in understanding the mechanisms by which sludge is converted into oil. In 1982, Bayer and Kutubuddin postulated that catalyzed vapor-phase reactions converted the lipids and proteins in sludge to straight-chain hydrocarbons. Analysis of the product oil confirmed that alkanes were produced, although pyrolysis processes invariably produce cyclic and aromatic compounds regardless of whether sludge, refuse, or cellulose is used as a substrate. It appears that the aluminosilicates and heavy metals in sludge catalyze the reaction, providing a silver lining to the problem of sludge contamination.[8]

11.3.1 Process

As in nature, dead cellular material is the principal substrate for oil production. The OFS process takes sludge that has been dried to 95% solids and heats it anaerobically to 350°C at essentially atmospheric pressure for about 30 min. In this first step roughly 40 to 50% of the sludge is vaporized. These vapors are contacted with the char (residue from the sludge) in the second stage of the reactor, where catalyzed vapor-phase reactions convert the organics to straight-chain hydrocarbons, the principal ingredient of crude oil. The process produces between 150 and 300 l of oil per ton of sludge processed, as well as char, a noncondensible gas, and reaction water. These by-products are combusted to produce most of the energy for sludge drying and reactor heating.

The oil yield from raw sludge is generally greater than that from digested sludge, and the oil from digested sludge is of lower viscosity than the oil from raw sludge. Under optimal conditions, oil yields of greater than 17% from anaerobically digested sludge and 46% for mixed raw sludge have been routinely achieved, and energy conversion efficiency greater than 95% is common. The capital costs of the system are similar to those of an incineration plant, but net operating costs are substantially less. The technology offers environmentally benign end products with heavy metals bound in the ash, pathogens destroyed, and air emissions minimized and controlled.[9]

Although the OFS technology has not been implemented in full scale, it has been successfully demonstrated at 1-tpd pilot facilities in both Australia and Canada. The technology is patented by Canadian Patents and Development Limited and marketed worldwide by Enersludge Inc., a joint venture company of Campbell Group (Australia) and SNC (Canada).[10]

11.3.2 Fate of Contaminants

The fate of heavy metals in the OFS reactor has been studied extensively. In one study, 16 controlled OFS reactor runs yielded the products shown in Table 11-2.[11] Heavy metals and organochlorine analyses were performed on all 16 runs to determine

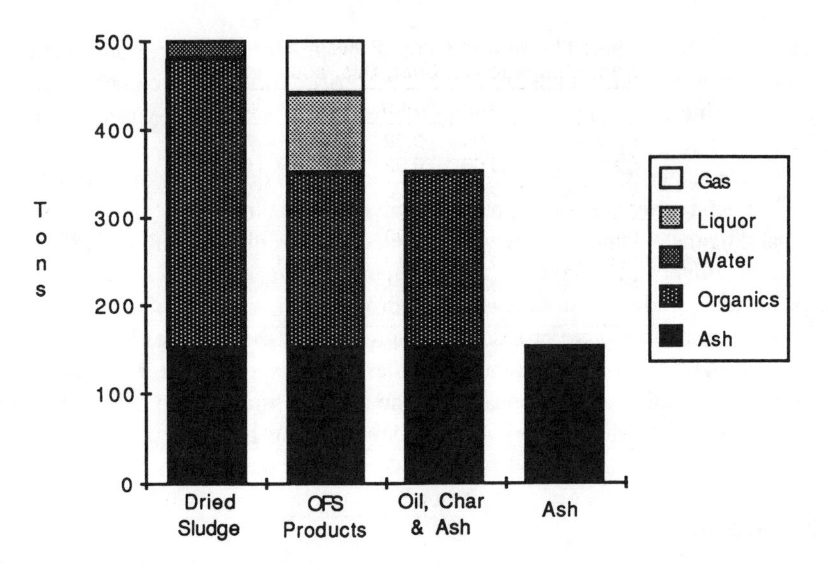

Figure 11-3 Volume reduction by oil from sludge (OFS) process. (Courtesy of Innotek Corporation, Little Rock, AR)

Table 11-2 OFS Products

Oil	14%
Char	58.5%
Noncondensible gas	11.5%
Reaction water	15.8%

Source: Bridle, Hammerton and Hertle.[11]

Table 11-3 Contaminant Concentrations in Sludge and Reactor Products

	Sludge		Oil		Char		Reaction water	
Contaminant	**Range**	**Average**	**Range**	**Average**	**Range**	**Average**	**Range**	**Average**
Arsenic	1–5	3.1	0.5–1	0.8	1–6	3.3	<1–3	2.2
Cadmium	35–54	44	0.1–1	0.7	59–87	69	<0.5	<0.5
Chromium	314–745	615	0.4–3.3	2.1	774–1080	977	<0.5	<0.5
Copper	1070–1380	1187	2–5	4	1615–2160	1934	<0.5	<0.5
Lead	245–310	267	0.3–8	1.1	358–460	412	<0.5	<0.5
Mercury	6–9	7	<1	<1	0.6–2	1.5	<1	<1
Nickel	207–332	198	0.4–1	0.9	258–350	307	<0.5	<0.5
Zinc	2655–3390	1999	0.7–13	8.7	2655–3365	3142	<0.5	<0.5

Source: Bridle, Hammerton and Hertle.[11]
Note: Units are mg/kg.

which of the OFS products contained most of the contaminants. The results of these tests are shown in Table 11-3. Multiplying the concentrations of the various heavy metals by the fraction of each constituent in the products gives the metals loadings of the char, oil, gas, and water, which can then be converted into fraction of total loadings. These calculations are shown in Table 11-4.

Table 11-4 Partitioning of Heavy Metals from Sludge to Oil, Char, Gas, and Water

	Oil	Char	Gas	Water
Arsenic	0.04	0.62	0.23	0.11
Cadmium	0.00	0.92	0.08	0.00
Chromium	0.00	0.93	0.07	0.00
Copper	0.00	0.95	0.05	0.00
Lead	0.00	0.90	0.10	0.00
Mercury	0.02	0.13	0.83	0.02
Nickel	0.00	0.91	0.09	0.00
Zinc	0.00	0.92	0.08	0.00

Source: Bridle, Hammerton and Hertle.[11]

Table 11-5 Fate of Heavy Metals in the Char Combustor

	Bed temperature 786°C		Bed temperature 876°C	
Metal	**% in ash**	**% in flue gas**	**% in ash**	**% in flue gas**
Arsenic	95	0	98	<0.14
Cadmium	78	<0.14	76	<0.15
Chromium	62	0	89	0
Copper	78	<0.05	78	0.05
Lead	92	0	88	<0.14
Mercury	5.4	2.6	1.7	5.3
Nickel	62	<0.3	69	<0.3
Zinc	69	<0.03	81	<0.03

Source: Bridle, Hammerton and Hertle.[11]

In a full-scale OFS plant the char, gas, and oil are combusted to provide the energy required to dry the sludge and heat the reactor. Pilot tests of char combustion show that excellent organic carbon efficiency can be achieved — 99.992 to 99.996%. More than 90% of the ash was retained in the bed as a granular product much like the char feed. At a bed temperature of 786°C, 59% of the char was converted to sulfur dioxide, while nitrogen oxides remained steady at 300 to 400 ppm. Mass balances of heavy metals during char combustion are shown in Table 11-5.

11.3.3 Costs

In 1984, Environment Canada contracted with Zenon Environmental Inc. of Burlington, Ontario to assess the costs of the OFS technology. A 25-t OFS plant was designed and the economics of the operation were compared to the cost of existing sludge incinerators. The sludge treatment train in each plant includes digestion, thickening, dewatering, conditioning, incineration, and ash disposal. Capital costs were taken from actual construction contracts and were calculated by amortizing the total capital costs over 20 years at 10% interest. Operation and maintenance costs were determined using plant records and interviews of plant personnel. At existing sludge streams of 36 to 67 t/day, the total costs were from $350 to $1042/t (Canadian dollars).[10]

Preliminary costs for the OFS technology were developed based on the data from bench- and pilot-scale studies. For a 45 t/day plant, the capital costs, including dewatering, conditioning, drying, conversion condensation, oil and water separation, char combustion, and ash disposal, were estimated at $17.5 million, plus $0.5 million for performance evaluation and environmental assessment. Based on an amortization of 11% per year over 20 years and an annual operating cost of $2.2 million/year, the total unit cost is $295/t dry solids, a cost which does not include the value of the oil produced.[10]

11.4 SLUDGE TO ENERGY — HYPERION ENERGY RECOVERY SYSTEM

In the early 1970s Los Angeles was faced with a court order to stop the ocean dumping of sludge by July 1985. The Los Angeles Hyperion treatment plant, which produces roughly 200 dry tons of anaerobically digested sludge a day, had initially intended to land apply the sludge on nearby desert in a project called "Desert Bloom". Intense public opposition ruled that out. By the late 1970s, the plan had evolved into land-applying sludge on an agency-owned farming operation, an idea that was also squashed by public disapproval. As a result, in the early 1980s Hyperion decided to process all sludge on-site through the Hyperion Energy Recovery System. (HERS). HERS is the latest generation of sludge incineration and has been called "the most comprehensive project in the world involving energy recovery from sludge". Rather than simply burning the sludge, HERS uses sludge as a fuel and generates more energy than it uses. One of the largest projects ever built by the City of Los Angeles Department of Public Works, the HERS system was completed in 1987.[12]

Every possible kilowatt is squeezed from the sludge through HERS. Between 1987 and 1992 over 450 million kWh of electrical power have been generated, with an average power production of about 20 MW. Most of the electricity is used to power the facilities at the Hyperion wastewater treatment plant, while the rest is sold to the local electric utility.[13]

11.4.1 Process

Each process train for sludge dehydration consists of a feed preparation system, four evaporation chambers, a heat exchanger, an oil separator, and a solids/oil separator system, as shown in Figure 11-4. A covered transport system moves sludge pellets to fuel storage bins before combustion. In the initial process design, anaerobically digested sludge was centrifuged to a solids content of 20 to 22%. The Carver-Greenfield multiple-effect evaporation process converts the sludge cake into pellets that are 99% solids.[13]

The electricity produced by HERS comes from four gas turbines run by the digester gas and two steam turbines run by steam. Gas turbines produce heat in the process of generating electricity; instead of venting this hot exhaust as waste heat it is fed to generators that produce steam, which is used, in turn, to generate electricity. This "cogeneration" approach nearly doubles the net production of electricity. A major advantage of using gas turbines is their low air emissions output.

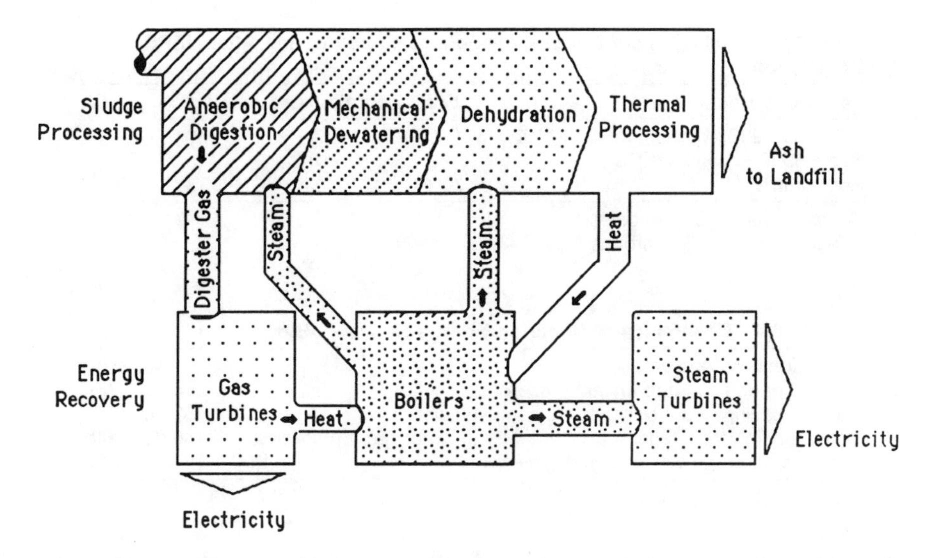

Figure 11-4 Hyperion Energy Recovery System: HERS. (Courtesy of City of Los Angeles Hyperion Treatment Plant, Playa del Rey, CA)

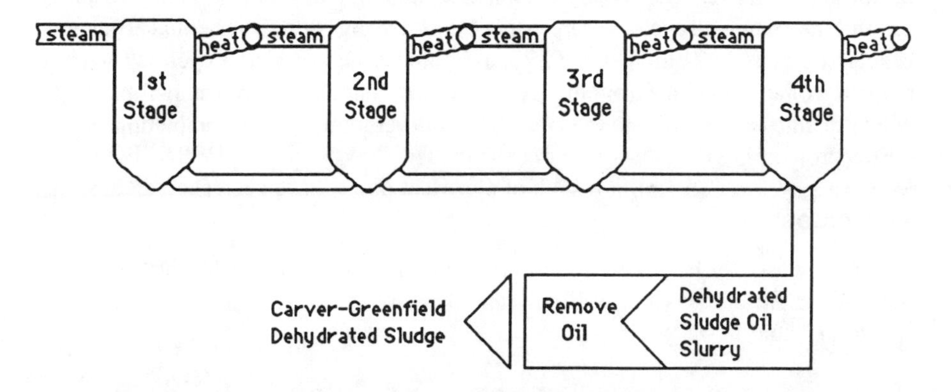

Figure 11-5 Carver-Greenfield multiple effect dehydration. (Courtesy of City of Los Angeles Hyperion Treatment Plant, Playa del Rey, CA)

In multiple-effect dehydration, dewatered sludge is placed inside chambers. When heat is applied to one chamber, water vapor is driven off. The heat of that water vapor is applied to the adjacent chamber, from which more vapor is boiled off. When this method is applied to four chambers in a row, as seen in Figure 11-5, the Carver-Greenfield dehydration system removes water in a very energy-efficient manner.[13] The efficiency of the sludge dehydration is the thermodynamic heart of HERS.

In Los Angeles, the air emissions of any industrial process are very tightly controlled. Air emissions from the Hyperion treatment plant have been substantially reduced by HERS operation. Reductions in all air pollutants are accomplished by retiring the internal combustion engines, while nitrogen oxides are reduced using

Table 11-6 Emissions of Primary Air Pollutants Before and After HERS Completion

Emissions (lb/day)	NO	SO	Particulates	NMHC[a]	CO
Hyperion 1982, without HERS	1800	660	225	650	1740
Hyperion 1987, with HERS	1075	165	110	45	780
Reduction	40%	75%	51%	93%	55%

Source: City of Los Angeles.[13]
[a] Nonmethane hydrocarbons.

Table 11-7 Economics of Hyperion Energy Recovery System

Total construction costs	$201,400,000
Annual operations and maintenance costs	$14,400,000
Annual savings from being energy self-sufficient	$9,100,000
Total funding from federal and state clean water grant programs	$140,000,000
Total costs to the City of Los Angeles	$61,400,000

Source: City of Los Angeles.[13]

fluid bed gasification and multistage combustion as well as water injection to the gas turbines. Bag houses reduce particulate levels; afterburners and catalytic converters take care of the CO (carbon monoxide) and nonmethane hydrocarbons (NMHCs). Sulfur oxides are reduced by wet chemical scrubbing and chemical addition, and a gas-turbine combined cycle power system cleans the digester gas. Thermal efficiency is approximately 70%, high even for today's state-of-the-art power cycles. The total reduction of air pollutants from the Hyperion treatment plant is shown in Table 11-6, which quantifies the emissions from Hyperion before and after completion of the HERS project.[13]

11.4.2 Costs

The expenses for HERS construction were shared by the U.S. EPA, the California State Water Resources Control Board, and the City of Los Angeles. Total costs are shown in Table 11-7.[13]

11.5 SLUDGE BRICKS AND TILES

The concept of using sludge in manufacturing ceramic products has historic roots dating back to at least 1889, when Thomas Shaw was issued an English patent for "Improvements in Utilizing the Waste Products from Sewage Works for the Manufacture of Bricks, Tiles, Quarries, Building Slabs and the like". After nearly a century, the technology was revived in this country in 1982 when Maryland Clay Products, Inc. used 20 tons of sludge from Bowie, MD in the production of 35,000 bricks. This was followed by a production run in which 120 tons from the Parkway plant were used in the production of 500,000 bricks that were used by the Washington Suburban Sanitary Commission (WSSC) for facilities construction in Maryland's Montgomery and Prince George's counties.

Explained Dr. Edward Bryan of the National Science Foundation in Washington, D.C.: "It was expected that the firing process to which the formed bricks were subjected would destroy any pathogenic bacteria or other organisms present in the sludges used and that any organic substances present, including those of toxic significance, would be completely destroyed. Heavy metals and nonvolatile substances were expected to become encapsulated in the ceramic matrix of the finished product, thus reducing or eliminating any further risk to health or environmental quality that is commonly associated with other methods of sludge management.... Finally, the concept appeared to be readily adaptable to current practice in manufacture of bricks."[14]

In Japan, the shortage of landfills has prompted interest in sludge recycling. Initially incineration was seen as the solution, and by 1988 more than 58% of Japan's sludge was incinerated. This increased the life of existing landfills, but did not provide a long-term solution, prompting the development of a three-pronged approach to sludge reuse: using sludge as agricultural compost, for green space enhancement, and as a source for construction material. The sludge melting system is a treatment process that generates construction material while reducing the volume and weight of the sludge and stabilizing the heavy metal content. It is rapidly gaining popularity in Japan, where 11 furnaces are in operation or under construction. The Fujimi Tile Company of Nagoya, Japan has patented a method for making ceramic bricks and tile out of sewage sludge, and concrete secondary products such as interlocking paving blocks and hume pipe are now made from concrete using sludge as aggregate. The sludge slag is also being tested for use as a road subbase. The benefits of using the sludge ash for construction materials are twofold: the material is diverted from landfills, and the metals from the sludge that have been transferred to the ash are immobilized. While sludge contaminants may migrate into the environment from landfill leachate, the slag produced by the sludge melting system effectively immobilizes heavy metals indefinitely.[15]

11.5.1 Process

In the sludge melting system, sludge is heated at temperatures between 1200 and 1500°C to evaporate the water, thermally decompose and burn off the organic components, and melt the inorganic components. At these high temperatures, efficient heat transfer and appropriate handling of the waste gas are critical. The entire treatment process involves five operations: pretreatment, melting, heat recovery, waste-gas treatment, and slag production, as shown in Figure 11-6.[15]

Pretreatment consists of drying the sludge with steam or hot air driers and incinerating the sludge. The water content of the resulting sludge ranges from 5 to 40%, depending on the type of drier used. The sludge is then mechanically crushed or granulated before introduction to the melting furnace.

Melting furnaces generate high-temperature combustion gas. Waste heat is recovered from the combustion gas to be used as the heat source for the pretreatment and melting processes. The heat is normally recovered by waste boilers as steam. Because of the highly calorific content of the coke, electricity can be generated during coke-bed melting with steam dynamos.

The shape and properties of the slag are affected by the cooling methods. Casting molten sludge in water produces a vitreous, fine-grained material with a smooth surface. Air-cooled slag is vitreous and massive, while reheating the slag to about

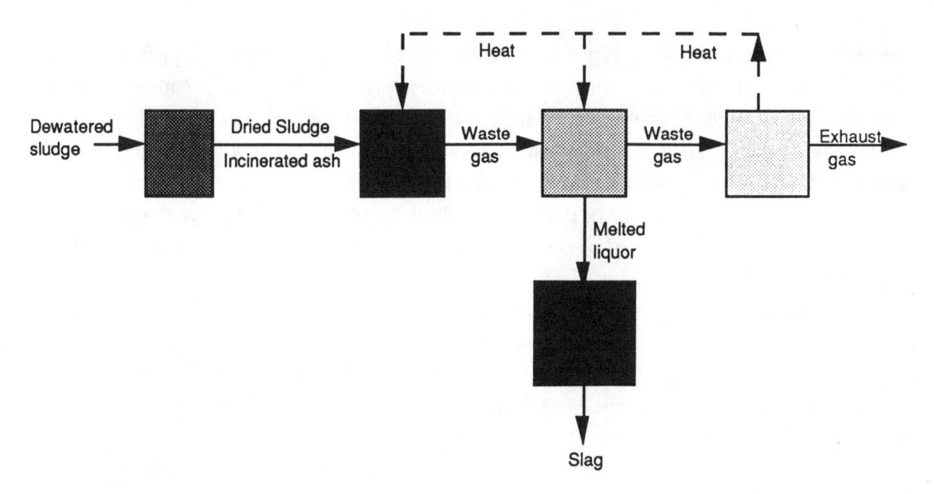

Figure 11-6 Sludge melting system. (From Oshima, Y. and Masuta, T., *Water Environ. & Technol.,* Jun. 1991. With permission.)

Table 11-8 Economics of Sludge Melting in Japan, 1991

Capacity (tons/day)	Pretreatment process	Melting furnace	Construction $/dry ton	Operating $/dry ton	Maintenance $/dry ton	Total costs $/dry ton
10	Steam dry	Coke bed	$0.76	$177	$100	$278
16.5	Carbonizing	Cyclone	$0.43	$46	$26	$72
25	Steam dry	Reverberatory	$0.45	$95	$19	$114
40	Steam dry	Coke bed	$0.35	$85	$10	$95

Source: Ohshima and Masuta.[15]

1000°C after air cooling produces crystallized slag. The compressive and bending strength of the crystallized slag is comparable to that of sand, with progressively lower results from air- and water-cooled slag.

Waste gas generated from the pretreatment driers and melting process furnaces is treated by a wet or dry electrostatic precipitator, a denitrizer, and a deodorizer. Most of the waste gas from the sludge dryer is recycled as a carrier gas for the dryer after dehumidification, while part of the gas is diverted for combustion to heat the melting furnace. The gas given off by the dryer circulates within a closed system, so none of it is discharged into the environment. However, the wastewater generated from dehumidification should be monitored, since the odor and dust carried in the gas are transferred to the wastewater.

After melting, molten inorganic materials are converted to products that can be beneficially reused. Methods of extracting the molten matter include placing it in water and cooling it in air. Crystallization, which produces higher quality slag, is being investigated.[16]

11.5.2 Costs

The sludge melting system is costly compared to most other methods of beneficial reuse because it takes pelletized sludge and adds a significant amount of processing energy. In Japan, where landfill pressures make the disposal of incinerator ash quite expensive, the sludge melting system becomes economically feasible. The Japanese Sewage Works Agency found the construction, operating, and maintenance costs per unit volume of sludge treated were inversely proportional to the size of the facility, as shown in Table 11-8. Sludge feedstock with lower water content and higher volatile solids content makes sludge melting less costly.[15]

REFERENCES

1. Case, F. "Waste Not," *The Seattle Times,* October 3, 1988, Section F1.
2. Molton, P.M., A.G. Fassbender, and M.D. Brown. "STORS: The Sludge-to-Oil Reactor System," EPA/600/S2-86/034, U.S. EPA, Cincinnati, OH (June 1986).
3. Coan, H. "'Poop Pump' Creates Diesel From Dumpings," *Arkansas Democrat,* October 20, 1988.
4. Stockton, W. "A New Use for Sludge: To Make Oil," *The New York Times,* December 16, 1987.
5. Donald, L. "Little Rock Company Finding Sewer Sludge Can Be Beautiful," *Arkansas Gazette,* October 23, 1988.
6. "STORS Process Mass Balance," Fact Sheets from Innotek Corporation, Little Rock, AR (1990).
7. Draeger, H. "Sanitary District Gets Sludge to Oil Proposal," *Chicago Sun Times,* September 23, 1988, p. 2
8. "Creating Fuel Oil from Sewage Sludge," Wastewater Technology Centre, Environment Canada, Burlington, Ontario (1988).
9. Bridle, T.R., C.K. Hertles, and T. Luceks. "The Oil from Sludge Technology: A Cost Effective Sludge Management Option," 13th Federal Convention of the Australian Water and Wastewater Association, Canberra, Australia (March 1989).
10. Campbell, H.W. and D.A. Martinoli. "Canada's Oil-From-Sludge Technology," *Water Environ. & Technol.* (July 1991).
11. Bridle, T.R., I. Hammerton, and C.K. Hertle, "Control of Heavy Metals and Organochlorides Using the Oil from Sludge Process," Enersludge (Australasia) Pty Lyd, West Perth, Australia and the Sydney Water Board, Surry Hills, NSW 2010, Australia.
12. Barletta, R.J. and R.A. Webber. "A Tale of Three Giant Sewerage Systems," *J. Water Pollut. Control Fed.* 58(9) (1986).
13. "HERS Project Update," City of Los Angeles Hyperion Energy Recovery System, September 1986.
14. "Constructing with Biobricks," *Biocycle* (February 1986).
15. Ohshima, Y. and T. Masuta. "Sludge Melting: Japan's Method of Treatment," *Water Environ. & Technol.* (June 1991).
16. Tay, J.H. and W.K. Yip. "Sludge Ash as Lightweight Concrete Material," *J. Environ. Eng.* 115(1) (1989).

12 REGULATORY LIMITATIONS

12.1 GENERAL ISSUES

The regulation of sludge application by federal and state government began with a 1977 amendment to Section 405 of the Clean Water Act (CWA), which required the EPA to develop regulations containing guidelines for the use and disposal of sewage sludge. These regulations, which were intended to cover the marketing and distribution of sludge and sludge products, were nearly completed in 1981, but work on them then stopped. In 1987, Congress amended Section 405 again. For the first time, a comprehensive program for reducing the potential environmental risks and maximizing the beneficial use of sludge was set forth, along with a timetable for development of sludge use and disposal guidelines. The agency was required to identify possible toxic pollutants in sludge and to promulgate regulations that specify acceptable management practices and numerical limitations for sludge that contains these pollutants.

Heavy metals or other elemental contaminants are ubiquitous in the environment, and the complete removal of metals from sewage sludge is not practicable. Instead, state and federal regulations establish maximum safe application rates and regulate the concentrations of toxicants that industrial contributors can introduce into the waste stream. The regulations had to be "adequate to protect human health and the environment from any reasonable anticipated adverse effect of each pollutant".

The final Part 503 rule (promulgated in May 1989 and revised in February 1993) covers three general categories of use and disposal practices: application of sewage sludge to land, including farms, gardens, forests, and reclamation sites; surface disposal on dedicated sites or in sewage-sludge-only landfills; and incineration at sewage-sludge-only incinerators. It does not cover sludge that is codisposed in a municipal solid waste landfill or incinerator (which is covered by 40 CFR 258). The standards apply to publicly and privately owned treatment works that generate or treat domestic sewage sludge, as well as to any person who uses or disposes of sewage sludge from these treatment works. It also includes new regulations for domestic septage applied to land or surface disposed.

For further details see the EPA 40 CFR Part 503; Robert Bastion's "Summary of 40 CFR Part 503 Standards for the Use or Disposal of Sewage Sludge," EPA unpublished memo (February 25, 1993, revised March 8, 1993); and from Nora Goldstein's "EPA Releases Final Sludge Management Rule," *Biocycle* (January 1993).

In most cases the rule is "self-implementing", which means that citizen lawsuits or the EPA can enforce regulation even before permits are issued. Compliance with the rule must be achieved, with some exceptions, by February 19, 1994 or 1 year later if construction of new pollution control facilities is required for compliance.

In the Part 503 rule, the EPA created standards for those pollutants for which it had sufficient information to establish protective numerical limits, management practices, and other requirements. This first stage is identified as Round One. The agency recognizes that the Part 503 rule may have missed some pollutants. Round Two, which will identify whether any additional sludge pollutants need regulation, has already begun.

If a state wishes authorization to run its own sludge program, the Part 503 regulations must be adopted as minimum requirements, and the program must be approved in accordance with 40 CFR Parts 122, 123, and 501. While the EPA would like the states to adopt the numerical criteria and other requirements in Part 503 without significant changes, Section 405 of the Clean Water Act provides that nothing in the section is intended to waive more stringent requirements in the CWA or any other law.

12.1.1 Permitting

All treatment works, including nondischargers and sludge-only facilities, must apply for a permit. Commercial handlers that only distribute or land apply the sludge are not required to obtain a permit unless specifically requested to do so be the permitting authority — the EPA or an approved state. The Part 503 standards will be incorporated into National Pollution Discharge Elimination System (NPDES) permits issued by the EPA or by states with an approved sewage sludge management program. In accordance with NPDES permit renewal procedures, permit applications must be submitted at least 180 days before their permit is due to expire. Permitting authorities may request that permit applications are submitted earlier, with permit applications being due 180 days after such a request.

12.1.2 Monitoring

Parameters which must be monitored include the metals arsenic (As), cadmium (Cd), chromium (Cr), copper (Cu), lead (Pb), mercury (Hg), molybdenum (Mo), nickel (Ni), selenium (Se), and zinc (Zn). Indicator organisms (fecal coliforms as the most probable number [MPN] *or* colony forming units [cfu] per gram total solids) or pathogens (*Salmonella* most probable number or colony forming units per 4 g total solids). Each sampling event for fecal coliforms must have seven separate samples, with the geometric mean of the results used to determine compliance with the standard. Vector attraction reduction, the degree of stabilization or odor abatement to prevent the attraction of rat, flies, and other vectors, is also monitored.

The minimum frequency of monitoring for metals, indicator organisms, and vector attraction reduction requirements shall be based on the amount of sewage sludge produced annually, as shown in Table 12-1, although more frequent monitoring

Table 12-1 Monitoring Frequency

Sewage sludge amounts (dry metric tons per year)	Monitoring frequency
>0 to <290	Once per year
290 to <1,500	Once per quarter
1,500 to <15,000	Once per 60 days
≥15,000	Once per month

requirements may be imposed by the permitting authority. After a period of 2 years the permitting authority may allow the monitoring frequency to be reduced to no less than once a year. In general, records must be kept for 5 years.

12.2 THE PART 503 REGULATIONS

12.2.1 Numerical Limits for Contaminants

40 CFR Part 503 required that the EPA develop a complete set of regulations that protect different environmental media equally. Limits were set for 10 pollutants, with 14 pathways of exposure evaluated for each pollutant. The regulations are based on the "highly exposed individual," the person, plant, or animal that is highly exposed to the possible effects of sludge applied contaminants, and the subpopulations at higher risk, such as children. The most restrictive number from the most limiting pathway was used to establish the cumulative limits.

Tables 12-2–12-5 list the pollutant limits for land application. To be land applied, bulk sewage sludge must meet the pollutant ceiling concentrations (Table 12-2) *and* either cumulative pollutant loading (Table 12-3) *or* pollutant concentration limits (Table 12-4). Bulk sewage sludge applied to lawns and home gardens must meet the pollutant concentration limits (Table 12-4). Sewage sludge sold or given away in bags must meet the pollutant concentration limits (Table 12-4) *or* annual sewage sludge product application rates that are based on the annual pollutant loading rates (Table 12-5).

12.2.2 "Exceptional Quality" Sludge

If a sewage sludge meets the pollutant concentration limits shown in Table 12-4, class A pathogen reduction requirements, and a vector attraction reduction processing option, it will usually be exempt from the management practices and general requirements applicable to land application practices. This innovative feature of the 40 CFR 503 regulations creates an incentive for POTWs to improve sludge quality before embarking on a land application program.

12.2.3 Pathogen Reduction Requirements

Municipal sewage sludge contains substantial populations of microorganisms which are typically found in the human gut. Some of these human enteric organisms

Table 12-2 Ceiling Concentration Limits

Pollutant	Ceiling concentration limits absolute values (mg/kg)
Arsenic	75
Cadmium	85
Chromium	3000
Copper	4300
Lead	840
Mercury	57
Molybdenum	75
Nickel	420
Selenium	100
Zinc	7500

Source: U.S. EPA.[8]

Table 12-3 Cumulative Pollutant Loading Rates

Pollutant	Cumulative pollutant loading rates (kg/ha)
Arsenic	41
Cadmium	39
Chromium	3000
Copper	1500
Lead	300
Mercury	17
Molybdenum	18
Nickel	420
Selenium	100
Zinc	2800

Source: U.S. EPA.[8]

Table 12-4 "High Quality" Pollutant Concentration Limits

Pollutant	"High quality" pollutant concentration limits, monthly averages (mg/kg)
Arsenic	41
Cadmium	39
Chromium	1200
Copper	1500
Lead	300
Mercury	17
Molybdenum	18
Nickel	420
Selenium	36
Zinc	2800

Source: U.S. EPA 40.[8]

can, through exposure to susceptible hosts under certain conditions, lead to illness in the host. As a result, the U.S. EPA requires that municipal sludges receive specific levels of pathogen reduction before land application to reduce the potential for impacts on human health.

The Part 503 regulations classify sewage sludge into two categories, class A and class B. All sewage sludge that is land applied must meet at least class B requirements. Bulk sewage sludge that is land applied to lawns and home gardens and sewage sludge that is sold or given away in bags or other containers must meet the class A criteria and one of the vector attraction reduction options.

12.2.3.1 Class A Sewage Sludge

Class A sludge must meet one of the following criteria:

1. A fecal coliform density of less than 1000 MPN/g total dry solids (TS);

or

2. A *Salmonella* sp. density of less than 3 MPN/4 g TS;

or

the requirements of **one** of the following alternatives must be met:

and

1. **Time/temperature** — An increased sewage sludge temperature should be maintained for a prescribed period of time according to the guidelines shown in Table 12-6;

or

2. **Alkaline treatment** — The pH of the sewage sludge is raised to greater than 12 for at least 72 h. During this time, the temperature of the sewage sludge shall be greater than 52°C for at least 12 h. In addition, after the 72-h period, the sewage sludge is to be air dried to at least 50% TS.

or

3. **Prior testing for enteric viruses/viable helminth ova** — If the sludge is analyzed *before* the pathogen reduction process and found to have densities of enteric virus <1 plaque forming unit (pfu) per 4 g TS and <1 viable helminth ova per 4 g TS, the sludge is class A with respect to enteric viruses and viable helminth ova until the next monitoring session. If the sludge is monitored *before* the pathogen reduction process and found to have densities of enteric virus ≥1 pfu/4 g TS or ≥1 viable helminth ova per 4 g TS and is tested again *after* processing and found to meet the enteric virus and viable helminth ova levels listed under point 4 below, then the sludge will be class A with respect to enteric viruses and viable helminth ova when the operating parameters for the pathogen reduction process are monitored and shown to be consistent with the values or ranges of values documented.
4. **No prior testing for enteric viruses/viable helminth ova** — If the sludge is not analyzed *before* pathogen reduction processing for enteric viruses and viable helminth ova, the sludge must meet the enteric virus and viable helminth ova levels noted below to be class A at the time the sludge is used or disposed of, prepared for sale, or given away in a bag or container, or when the sewage sludge or derived material meets "exceptional quality" requirements — pollutant concentration limits, class A pathogen reduction, and vector attraction reduction requirements:

   - The density of enteric viruses must be <1 pfu/4 g TS.
   - The density of viable helminth ova must be <1 per 4 g TS.

or

5,6. The sewage sludge is treated by a PFRP or a PFRP-equivalent process, listed in Table 12-7.

Table 12-5 Annual Pollutant Loading Rates

Pollutant	Annual pollutant loading rates (kg/ha/year)
Arsenic	2.0
Cadmium	1.9
Chromium	150
Copper	75
Lead	15
Mercury	0.85
Molybdenum	0.90
Nickel	21
Selenium	5.0
Zinc	140

Source: U.S. EPA 40.[8]

Table 12-6 Time and Temperature Guidelines

Total solids	Temp (t)	Time (D)	Equation	Notes
≥7%	≥50°C	≥20 min	$D = \frac{131,700,000}{10^{0.14t}}$	No heating of small particles by warmed gases or immisicble liquid
≥7%	≥50°C	≥15 s	$D = \frac{131,700,000}{10^{0.14t}}$	Small particles heated by warmed immiscible liquid
<7%	>50°C	≥15 s to <30 min	$D = \frac{131,700,000}{10^{0.14t}}$	
<7%	≥50°C	≥30 min	$D = \frac{50,070.000}{10^{0.14t}}$	

Note: In no case would temperatures calculated using the appropriate equation be less than 50°C.

12.2.3.2 Class B Sewage Sludge

Class B sludge must meet one of the following pathogen reduction requirements:

1. The sewage sludge must be treated with a PSRP or a PSRP-equivalent process, as shown on Table 12-8;

or

2. At least seven sewage sludge samples should be collected at the time of use or disposal and analyzed for fecal coliforms during each monitoring period. The geometric mean of the densities of these samples will be calculated and should meet the following criteria:

- <2 million MPN/g TS

or

- <2 million cfu/g TS

In addition, for any land-applied sludge that meets class B pathogen reduction requirements, but not those of class A, additional site restrictions shall be observed.

12.2.4 Vector Attraction Reduction Requirements

Vector attraction reduction reduces the potential for spreading infectious disease agents by vectors such as flies, rodents, and birds. The alternatives for meeting the vector reduction requirements are as follows:

1. **Aerobic or anaerobic digestion** — Mass of volatile solids reduced by 38% or more. Volatile solids reduction is measured between the raw sludge before stabilization and the sewage sludge ready for use or disposal. This criterion should be easily met by properly designed and operating anaerobic digesters, but not as easily by typical aerobic digesters. POTWs with aerobic digesters may need to meet vector attraction reduction requirements through alternative 3 or 4.
2. **Anaerobic digestion** — If 38% volatile solids reduction cannot be achieved, vector attraction reduction can be demonstrated by further digesting a portion of the digested sewage sludge in a bench-scale unit for an additional 40 days at 30 to 37°C or higher and achieving a further volatile solids reduction of <17%.
3. **Aerobic digestion** — If 38% volatile solids reduction cannot be achieved, vector attraction reduction can be demonstrated by further digesting a portion of the digested sewage sludge with a solids content of ≤2% in a bench-scale unit for an additional 30 days at 20°C and achieving a further volatile solids reduction of <15%.
4. **Aerobic digestion** — Specific oxygen uptake rate (SOUR) is less than or equal to ≤1.5 mg O_2 per hour per gram of total solids at 20°C. If the SOUR criteria cannot be met, POTWs may be able to satisfy alternative 3.
5. **Aerobic processes** (e.g., composting) — Temperature is kept at >40°C for at least 14 days, and the average temperature during this period is >45°C.
6. **Alkaline stabilization** — The pH is raised to at least 12 by alkali addition and, without the addition of more alkali, remains at 12 or higher for 2 h and then at 11.5 or higher for an additional 22 h.

7,8. **Drying** — Total solids is at least 75% when the sewage sludge does not contain unstabilized primary solids and at least 90% when unstabilized primary solids are included. Blending with other materials is not allowed to achieve total solids percentage.

9. **Injection** — Liquid sewage sludge (or domestic septage) is injected beneath the surface, with no significant amount of sludge present on the surface after 1 h, except for sludges that are class A in pathogen reduction, which shall be injected within 8 h of discharge from the pathogen reduction process. This alternative is applicable to sludge that is land applied to agricultural land, forest, public contact sites, or reclamation sites; domestic septage that is applied to agricultural land, forest, or reclamation sites; and sewage sludge or domestic septage placed in a surface disposal site.
10. **Incorporation** — Sewage sludge (or domestic septage) that is land applied or placed in a surface disposal site shall be incorporated into the soil within 6 h of application, except for sludge that is class A for pathogen reduction, which, when land applied, shall be incorporated into the soil within 8 h of discharge from the pathogen reduction process. This alternative is applicable to bulk sewage sludge that is land applied to agricultural land, forest, public contact sites, or reclamation sites; domestic septage that is land applied to agricultural land, forest, or reclamation sites; and sewage sludge or domestic septage placed in a surface disposal site.
11. **Surface disposal daily cover** — Sewage sludge or domestic septage placed in surface disposal sites shall be covered with soil or other material at the end of each operating day.
12. **Domestic septage treatment** — The pH of domestic septage is raised to 12 or higher by alkali addition and, without the addition of more alkali, remains at 12 or higher for 30 min. This alternative is applicable to domestic septage applied to agricultural land, forest, or reclamation sites or placed in a surface disposal site.

When bulk sewage sludge is applied to agricultural land, forests, public contact, or reclamation sites, one of the vector reduction alternatives 1 through 10 must be met. When bulk sewage sludge is applied to home gardens or when sewage sludge is sold or given away in bags or other containers for land application, one of the alternatives 1 through 8 must be met. When sewage sludge is placed in a surface disposal site, one of the alternatives 1 through 11 must be met. When domestic septage is handled separately from sewage sludge, one of alternatives 9, 10, or 12 must be met when it is applied to agricultural land, forest, or reclamation sites, and one of alternatives 9 through 12 must be met when it is placed in a surface disposal site.

Table 12-7 Pathogen Treatment Processes: Processes to Further Reduce Pathogens (PFRP)

1. *Composting* — Using either in-vessel or static aerated pile composting, the temperature of the sewage sludge is maintained at 55°C or higher for 3 days. Using windrow composting, the temperature of the sewage sludge is maintained at 55°C or higher for 15 days or longer. During this period, a minimum of five windrow turnings are required.
2. *Heat drying* — Sewage sludge is dried by indirect or direct contact with hot gases to reduce the moisture content of the sludge to 10% or lower. Either the temperature of the gas in contact with the sludge exceeds 80°C or the wet bulb temperature of the gas in contact with the sludge as the sludge leaves the dryer exceeds 80°C.
3. *Heat treatment* — Liquid sludge is heated to a temperature of 180°C or higher for 30 min.
4. *Thermophilic aerobic digestion* — Liquid dewatered sludge is agitated with air or oxygen to maintain aerobic conditions, and the mean cell residence time for the sewage sludge is 10 days at 55 to 60°C.
5. *Beta ray irradiation* — Sewage sludge is irradiated with beta rays from an accelerator at dosage of at least 1.0 Mrad at room temperature (ca. 20°C).
6. *Gamma ray irradiation* — Sewage sludge is irradiated with gamma rays from certain isotopes such as ^{60}Co and ^{137}Ce, at dosages of at least 1.0 Mrad at room temperature (ca. 20°C).
7. *Pasteurization* — The temperature of the sludge is maintained at 70°C or higher for at least 30 min.

Table 12-8 Pathogen Treatment Processes: Processes to Significantly Reduce Pathogens (PSRP)

1. *Aerobic digestion* — Sewage sludge is agitated with air and oxygen to maintain aerobic conditions for a mean cell residence time and temperature between 40 days at 20°C and 60 days at 15°C.
2. *Air drying* — Sludge is dried on sand beds or on paved or unpaved basins for a minimum of 3 months; during 2 of the 3 months, the ambient average daily temperature is above 0°C.
3. *Anaerobic digestion* — Sludge is treated in the absence of air for a mean cell residence time and temperature of between 15 days at 35 to 55°C and 60 days at 20°C.
4. *Composting* — Using either in-vessel, static aerated pile, or windrow composting methods, the temperature of the sludge is raised to 40°C or higher for 5 days. For 4 h during the 5 days, the pile temperature must exceed 55°C.
5. *Lime stabilization* — Sufficient lime is added to the sludge to raise the pH of the sludge to 12 after 2 h of contact.

12.2.5 Management Practices for Land Application

Unless the sludge is "exceptional quality", the following management practices must be used for land application:

1. Bulk sewage sludge shall not be applied to flooded, frozen, or snow-covered ground unless authorized by the permitting authority.
2. Bulk sewage sludge shall not be applied at rates above agronomic rates, except for reclamation projects authorized by the permitting authority.
3. Bulk sewage sludge shall not be applied if it is likely to adversely affect a threatened or endangered species.
4. Bulk sewage sludge shall not be applied less than 10 m from waters of the U.S. unless authorized by the permitting authority.
5. Sewage sludge sold or given away shall have either a label affixed to the bag or container or an information sheet provided to the person who receives the sewage sludge for land application that provides information on proper use, including annual sludge application rates that conform with annual pollutant loading rates.

If sludge meets class B pathogen reduction rates, but not those of class A, and is land applied, the following site restrictions must be met:

1. Food crops with harvested parts that touch the sludge/soil mixture (such as melons, cucumber, squash, etc.) shall not be harvested for 14 months after application.
2. Food crops with harvested parts below the soil surface (root crops such as potatoes, carrots, and radishes) shall not be harvested for 20 months after application if the sewage sludge is not incorporated into the soil for at least 4 months.
3. Food crops with harvested parts below the soil surface (root crops) shall not be harvested for 38 months after application if the sewage sludge is incorporated into the soil in less than 4 months.
4. Food crops, feed crops, and other fiber crops shall not be harvested for 30 days after sewage sludge application.
5. Animals shall not be grazed on the site for 30 days after sewage sludge application.
6. Turf shall not be harvested for 1 year after sewage sludge application if the turf is placed on land with a high potential for public exposure or a lawn unless otherwise specified by the permitting authority.
7. Public access to land with high potential for public exposure shall be restricted for 1 year after sewage sludge application.
8. Public access to land with low potential for public exposure shall be restricted for 30 days after sewage sludge application.

12.3 RESEARCH ON THE FATE OF HEAVY METALS

There exists a vast body of research on the fate of heavy metals in soils that have been amended with sewage sludge. The metals studied most intensely are those that play no demonstrated role as plant nutrients or that cause disease after inhalation or ingestion. Metals that are toxic to humans, but not phytotoxic, where healthy crops can produce adverse reactions after human consumption, are of particular interest to sludge application researchers.

The concept of the "soil-plant barrier" was first introduced in 1980 for consideration of potential toxicity to the food chain if trace elements are applied to the soil. A first group of metals includes those that are so strongly adsorbed to the soil or in plant roots that they are not translocated into edible plant parts. For example, lead is so insoluble and mercury is so bound inside the fibrous plant roots that they don't get to the edible plant parts. Of the elements that can actually be taken up by plants from sludge-treated soils, some can reduce the yield or even kill crops. This phytotoxicity was noted when sludge quality was poorly regulated, stunting crops wherever strongly acidic soils, metal-sensitive crops, and sludges with high concentrations of zinc, copper, or nickel were combined.

12.3.1 Cadmium

Cadmium may be the most widely investigated of the metals found in sewage sludge. Although it accumulates in humans and other animals, cadmium does not seem to play an essential role in plant or animal nutrition. Sources of cadmium include smelting operations, automobile tires, and, ironically, superphosphate fertilizers. As with all metals, cadmium is found in the ambient environment in measurable quantities.

The quantity of cadmium consumed by citizens in the U.S. averages about 20 μg/day, with the recommended maximum daily intake of dietary cadmium being 70 μg/day.[1] Heavy cigarette smoking can contribute as much as 75 μg/day, the water supply may be a substantial contributor, and air pollution may be another contributor to the cadmium total. Cadmium accumulates in the kidney slowly over one's lifetime. In the U.S., we average about one tenth as much cadmium in our kidneys at age 50 (peak cadmium levels) as would be required to cause the first indication of possible injury to the most sensitive part of the population. In addition, soil cadmium is virtually unavailable if zinc concentrations are high. Zinc concentrations in sludge are typically far higher than cadmium concentrations and appear to result in a near total lack of bioavailability of cadmium.[2,3]

12.3.2 Lead

After cadmium, lead is probably the most studied nonessential element contained in sewage sludges. This cumulative poison is more toxic than cadmium. Relatively insoluble in soils, lead is generally quickly immobilized by adsorption and precipitation soon after it has been released to the environment.[4] A significant amount of lead has accumulated in the soil, particularly on land that is near public roadways — the legacy of a half century of automobiles run on leaded gasoline. Individuals who regularly consume wild game are risking lead poisoning because of the vast quantities of lead shot that is scattered into the environment each year.

Although lead is relatively insoluble in soil, agricultural crops grown on leaded soil will contain traces of this element. It has been estimated that over one half of the lead in the bodies of U.S. citizens is attributable to the food supply. The other major source of lead to the individual is tap water and the metal pipes that carry the water from the municipal mains to the faucet.

The bioavailability of lead is dependent on a variety of factors. If humans ingest soluble lead acetate during fasting, about 80% of the dose is absorbed. If the soluble lead is given along with a complete meal, they absorb less than 5% of the dose, with some individuals absorbing as little as 1%.[1]

12.3.3 Mercury

In the past, mercury has been released to the environment in enormous amounts. The horrors of Japan's Minimata mercury poisoning of 1959 resulted in the virtual ban of industrial discharge of mercury into the waterways. The environmental loading of mercury has been considerably reduced, although it still makes its way into our water and food and, therefore, our sludge. In areas subjected to acid rain, the levels of mercury released from the soil may be astonishingly high. Mercury is an essential element in batteries and thermometers, both of which end up in landfills, where the mercury-laden leachate may contaminate drinking water supplies and agricultural groundwater. Roadside levels of mercury in the soil remain elevated from a half century of fallout from leaded gasoline, and agricultural crops in these areas may be affected. In areas of the country where the drinking water is slightly acidic, the largest source of mercury to the sludge may be from the solder that fastens the joints of copper pipes that carry domestic drinking water.

Like cadmium, mercury is not concentrated in the muscles or milk of food animals, and while most livestock ingest some mercury through the consumption of grasses this does not have much impact on the food supply.[5] While acute and chronic toxicity in humans from mercury exposure has been well documented, the land application of sludge does not appear to be a significant pathway for this element.

12.3.4 Chromium

The major users of chromium include the leather tanning and steel manufacturing industries. An essential element, its increase in the human diet may lead to a decrease in diabetes. Trivalent chromium is poorly absorbed by animals, and only 0.5 to 1% of the 5 to 100 μg/day ingested by humans is retained.[6] Hexavalent chromium is a different story; it is toxic to plants as well as humans and other animals. Digested sludges usually contain little if any hexavalent chromium, since the digestion process generally reduces this species to the trivalent state.

Organically complexed chromium in sludge may behave in a fashion similar to sludge organic nitrogen. When applied, only a fraction of the total nitrogen — and chromium — is available. As the organic matter decomposes, both nitrogen and chromium are released and taken up by the established vegetation. When these elements are not taken up by plants, or transformed into other elemental soils which can bind to the soils, there is the potential for them to move down the soil column into the groundwater. While this is theoretically possible, practical experience shows that the land application of sludge does not increase the levels of soil chromium below the depth of incorporation, nor does it increase the level of plant chromium after the first crop year.[4]

12.3.5 Copper, Nickel, and Zinc

Copper, nickel, and zinc are essential nutrients for both plants and animals. Animal feedstocks are usually supplemented with these elements to ensure the good health of the herds, and regulatory control of these elements balances the benefits of use with the potential impacts of misuse. Plants exhibit toxicity symptoms for copper at some level above 20 ppm in the plant tissues. Although phytotoxicity occurs above those levels, humans do not seem to have a toxic reaction to copper. In fact, high levels of dietary copper are beneficial for poultry and swine.[4]

Nickel is unlikely to be taken up by plants in quantities sufficient to cause toxicity provided that the soil pH is kept reasonably high. There does not appear to be a retention mechanism or storage site in the human body, and chronic nickel toxicity in humans has not been noted. The metals most commonly absorbed and concentrated by plants and animals are copper and zinc, which are considered to be necessary micronutrients for plants and animals. Cadmium, nickel, and lead apparently are not normally absorbed by plants in significant amounts.

Zinc functions in a number of enzyme systems, and problems associated with zinc application are invariably related to deficiency rather than toxicity. High levels of dietary zinc appear to reduce the toxicity of other metals such as cadmium and copper.[7] One of the primary causes of zinc deficiency in our diets is the substitution of plant proteins for animal proteins, since zinc is far less available to animals from plant proteins than from animal proteins. The low availability of the zinc in plants to animals that consume these plants suggests that zinc has minimal potential to cause toxicity problems.

REFERENCES

1. Chaney, R.L. "Public Health and Sludge Utilization," *Biocycle* (October 1990).
2. U.S. Environmental Protection Agency. "Background Document: Cumulative Cadmium Application Rates," Docket 4004, EPA, Washington, D.C. (1979).
3. Chaney, R.L., R.J.F. Bruins, D.E. Baker, R.F. Korcak, J.E. Smith, Jr., and D.W. Cole. "Transfer of Sludge-Applied Trace Elements to the Foodchain," in *Land Application of Sludge — Food Chain Implications,* A.L. Page, T.J. Logan, and J.A. Ryan, Eds., (Chelsea, MI: Lewis Publishers Inc.) (1987), pp. 67–99.
4. Baird, R. and S.M. Gabrielian. "Trace Organics and Inorganics in Distribution and Marketing Municipal Sludges," EPA/600 S-1+88/001 (1988).
5. Cappon, C.J. "Content and Chemical form of Mercury and Selenium in Soil, Sludge and Fertilizer Materials," *Water Air Soil Pollut.* 22:95–104 (1984).
6. Bray, B.J., R.H. Dowdy, R.D. Goodrich, and D.E. Pamp. "Trace Metal Accumulations in Tissues of Goats Fed Silage Produced on Sewage Sludge Amended Soils," *J. Environ. Qual.* 14:114–118 (1985).
7. Bidwell, A.M. and R.H. Dowdy. "Cadmium and Zinc Availability to Corn Following Termination of Sewage Sludge Applications," *J. Environ. Qual.* 16:438–442 (1987).
8. "40 CFR Part 503," U.S. Environmental Protection Agency (1993).

Index

D

E

F

G

H

I

J

L

M

N

O

P

Q